Clean Solvents

ACS SYMPOSIUM SERIES **819**

Clean Solvents

Alternative Media for Chemical Reactions and Processing

Martin A. Abraham, Editor
University of Toledo

Luc Moens, Editor
National Renewable Energy Laboratory

American Chemical Society, Washington, DC

Library of Congress Cataloging-in-Publication Data

Clean solvents : alternative media for chemical reactions and processing / Martin A. Abraham, editor, Luc Moens, editor.

p. cm.—(ACS symposium series ; 819)

Includes bibliographical references and index.

ISBN 0–8412–3779–4

1. Solvents.. 2. Environmental chemistry— Industrial applications.

I. Abraham, Martin A., 1961- II. Moens, Luc, 1957- III. American Chemical Society. Division of Industrial and Engineering Chemistry. IV. American Chemical Society. Meeting (220th : 2000 : Washington, D.C.) V. Series.

TP247.5 .C47   2002
660′.29482—dc21   2001056651

The paper used in this publication meets the minimum requirements of American National Standard for Information Sciences—Permanence of Paper for Printed Library Materials, ANSI Z39.48–1984.

Copyright © 2002 American Chemical Society

Distributed by Oxford University Press

PRINTED IN THE UNITED STATES OF AMERICA

Foreword

The ACS Symposium Series was first published in 1974 to provide a mechanism for publishing symposia quickly in book form. The purpose of the series is to publish timely, comprehensive books developed from ACS sponsored symposia based on current scientific research. Occasion-ally, books are developed from symposia sponsored by other organizations when the topic is of keen interest to the chemistry audience.

Before agreeing to publish a book, the proposed table of contents is reviewed for appropriate and comprehensive coverage and for interest to the audience. Some papers may be excluded to better focus the book; others may be added to provide comprehensiveness. When appropriate, overview or introductory chapters are added. Drafts of chapters are peer-reviewed prior to final acceptance or rejection, and manuscripts are prepared in camera-ready format.

As a rule, only original research papers and original review papers are included in the volumes. Verbatim reproductions of previously published papers are not accepted.

ACS Books Department

Contents

Indexes

Preface

During the past several years, the concept of green chemistry has taken hold within the academic community and has begun to move into the industrial sector. One of the 12 principles of green chemistry is "The use of auxiliary substances (e.g., solvents, separation agents, and so on) should be made unnecessary wherever possible and innocuous when used." The development of novel chemistries and processes that rely on benign solvents clearly falls within this context. It was in this spirit that we organized a two-day Symposium titled "Clean Solvents" as part of the American Chemical Society (ACS) National Meeting in Washington D.C. in August 2000. The symposia held at the meeting marked the first substantial programming effort of the Green Chemistry and Engineering Subdivision of the ACS Division of Industrial and Engineering Chemistry, Inc.

The development of this book provided a permanent record for many of the papers that were initially presented during the Washington symposium. This Symposium brought together leaders in solvent research with a focus on topics such as ionic liquids, supercritical fluids, aqueous and fluorous media, and biomass-derived solvents. In addition, we evaluated numerous papers that were presented at the American Institute of Chemical Engineers (AIChE) Annual Meeting held in San Francisco, California in November 2000 for additional papers that were appropriate to the proposed volume. The chapters thus represent a rather broad overview describing the use of benign solvents as alternative solvent media.

We believe this is the first volume to cover the wide range of clean solvents that can be used for industrial processes, and the first to describe the benefits of these alternative solvents in the context of conventional technologies. Our goal was to produce a volume comprising chapters describing frontier chemistry that can serve as a starting point for educating researchers who are new to the field. Although previous volumes have focused on an individual solvent system, we are unaware of any books that discuss all of the clean solvent systems in a single volume. Moreover, the design of solvents and techniques for choosing among solvent alternatives is also relevant to the development of processes that take advantage of the concepts of green chemistry. For many of the solvent systems discussed, this volume provides new information that extends the state of knowledge beyond that which has been known by investigators over the past several years.

Because of the breadth of the volume, we expect that this book will be of wide interest to chemists and engineers who are evaluating alternative solvent systems in which to conduct chemical syntheses. Academic researchers will benefit from the collections of papers within their particular focus area and will gain an appreciation for the alternatives that could be considered. Industrial researchers will appreciate the practicality of the sections on solvent selection and industrial uses of novel solvents and will benefit from the extensive coverage of different solvent systems. For all readers, we believe that this volume will be a first-of-its-kind collection of chapters that cover the broad range of research developments in alternative solvents. As such, the proposed volume should have broad appeal in the scientific and technical community.

Acknowledgments

We thank all those who have contributed to bringing about this volume. This includes all of the authors who have contributed their papers to this volume and the reviewers who worked hard to ensure the quality of the contributions. In addition, we acknowledge the contributions of the leaders of the Green Chemistry and Engineering Subdivision; particularly Mike Matthews, the subdivision Chair, and Robin Rogers, the ACS Division of Industrial and Engineering Chemistry, Inc. Chair, who helped to spur the formation of the subdivision. We also thank the ACS Division of Industrial and Engineering Chemistry, Inc., who gave us the opportunity to organize these sessions; Kelly Dennis, our acquisitions editor, and Margaret Brown, the editorial and production manager, of the ACS Books Department, who patiently resigned themselves to our delays and missed deadlines; and our secretarial staff and colleagues at the University of Toledo and National Renewable Energy Laboratory.

Martin A. Abraham

Department of Chemical and Environmental Engineering
University of Toledo
3055 Nitschke Hall
Toledo, OH 43606-3390

Luc Moens

National Renewable Energy Laboratory
Center for Chemistry of BioEnergy Systems
1617 Cole Boulevard
Golden, CO 80401-3393

Chapter 1

Green Chemistry as Applied to Solvents

Paul T. Anastas[1]

White House Office of Science and Technology Policy, 424 Old Executive Office Building, Washington, DC 20502
[1]Visiting professor: Chemistry Department, University of Nottingham, University Park, Nottingham NG7 2RD, United Kingdom

Green Chemistry is a relatively new approach to making products and processes as environmentally benign as they can be. There are fewer areas of more importance in this endeavor than that of solvents. With solvents being of extremely high volume and very broad breadth of applicability, their potential for negative impact on human health and the environment is very large. Through the use of innovative Green Chemistry methodologies and techniques, the area of clean solvents is emerging as one of the most active and innovative areas of research and development in all of Green Chemistry and Engineering.

Green Chemistry and Engineering is being increasingly cited for its rapid growth and its effectiveness in meeting economic and environmental goals simultaneously.(1,2) Since the inception of this focus area in 1991 (3), one of the prime areas of research has been on the design and development of cleaner solvents. To understand the techniques being brought to bear on the area of cleaner solvents, it is useful to understand the background of Green Chemistry and Engineering.

The term "Green Chemistry" has been defined as:

"The design of chemical products and processes to reduce or to eliminate the use and generation of hazardous substances". (4)

While this short definition appears straight forward, it marks a significant departure from the manner in which environmental issues were, or weren't, considered in the up-front design of the molecules and molecular transformations that are at the heart of the chemical enterprise. [Note: A parallel definition of Green Engineering can be found in Reference 5]

In reviewing the definition of Green Chemistry, one of the most important elements is the concept of *design*. By requiring that the impacts of chemical products and chemical processes are included as design criteria, the definition of Green Chemistry has linked hazard considerations to performance criteria in an inextricable way.

A further aspect of the definition of Green Chemistry is found in the phrase *use and generation*. Rather than focusing only on the waste substances that may be inadvertently produced in a process, Green Chemistry also includes all substances that are part of the entire life-cycle (6). Therefore, Green Chemistry is more than a tool for waste minimization and efficiency optimization, although clearly both of these goals are accomplished through the use of Green Chemistry. Green Chemistry also incorporates the significant consequences of the use of hazardous substances ranging from regulatory, handling and transport, and liability issues to name a few. To limit the definition to address only waste would be to look at a single piece of a complex problem. As will be seen later, Green Chemistry is applicable to all aspects of the product life cycle.

The final term of note used in the definition of Green Chemistry is the word *hazardous*. It is important to note that Green Chemistry is a way of addressing risk reduction and pollution prevention by reducing the intrinsic hazards of the substances rather than the circumstances and conditions of their use that may affect their risk. To understand why it is important that Green Chemistry is a hazard based approach we have to revisit the concept of risk.

Risk in its most fundamental form is the product of hazard and exposure:

$$Risk = Hazard \times Exposure$$

A substance of some known and quantifiable hazard together with a quantifiable exposure to that hazard will allow us to calculate the risk associated with that substance. Virtually all common approaches to risk reduction have focused on reducing exposure to hazardous substances. Regulations have often required reduction in exposure through control technologies, treatment technologies, and personal protective gear such as respirators, gloves etc., in order to reduce risk.

By achieving risk reduction through hazard reduction, Green Chemistry addresses concerns for cost and potential for failure of exposure controls. Regardless of the type of exposure control ranging from engineering controls through personal protective gear, there will always be an up-front capital cost; to what degree this cost can be recouped will be situation specific but it will always be there. In contrast, there is no up-front capital cost necessarily associated with the implementation of Green Chemistry. While some Green Chemistry options may require capital investment, others may actually lower total cost of operations from the outset. This is frequently the case in some of the easiest to implement Green Chemistry technologies.

Exposure controls, because they rely on either equipment or human activity to accomplish their goals, are capable of failing. Respirators can rupture, air scrubbers can break down, etc. In cases where this occurs, risk goes to a maximum because the exposure is now occurring to a constant hazard. Green Chemistry, in contrast, does not rely on equipment, human activity or circumstances of use, but instead it involves changing the intrinsic hazard properties of the chemical products and transformations. Therefore, Green Chemistry does not have the same vulnerability to failure as does the traditional approach to hazard control.

Another important point must be made about the use of the term *hazard* in the definition of Green Chemistry. The types of hazards covered by this term are not limited to physical hazards such as explosivity, flamability, and corrosivity. The term certainly includes the full range of toxicity including acute, chronic, carcinogenicity and ecological toxicity. Hazards, however, must also include for the purposes of this definition, global threats such as global warming, stratospheric ozone depletion, resource depletion and bioaccumulating and persistent chemicals. To include this broad perspective in the definition is both philosophically and pragmatically consistent. It would certainly be unreasonable to address only some subset of hazards while ignoring or not addressing others. But more importantly, the intrinsic hazardous properties are the types of issues that can only be addressed through the proper design or redesign of chemistry and chemicals.

Green Chemistry

The adoption of Green Chemistry as one of the primary methods of pollution prevention is a fairly recent phenomenon. It is certainly reasonable to question why this fairly straight forward approach is only now taking hold. The answer is found in a combination of factors including economic, regulatory, scientific and even social. Each of these drivers have combined to make the 1990's the decade where Green Chemistry was introduced, implemented, and commercialized on a wide industrial scale.

Since the early 1960's, environmental statutes and regulations have proliferated at an exponential rate. With these regulations came cost, restrictions on the use of chemicals and increased testing of chemical substances to determine their hazard. The increased cost and restrictions on the use of certain chemicals provided powerful incentives to industry to find replacements, substitutes or alternatives. The toxicity testing required by many of these statutes provided a new knowledge, and awareness about the type and degree of the hazardous nature of many chemicals. As the collective knowledge resulting from this testing of chemicals began to grow in the scientific and industrial communities, there was a growing demand by the public to find out more about the chemicals that are present in their communities. This culminated in the 1980's with the passage of the Emergency Planning and Community Right-to-know Act (EPCRA), which provided for the public release of data on chemicals being released to the air, water and land by industry. The consequence for industry is that there has been a tremendous pressure on industry to not only reduce the release of toxic chemicals to the environment but to reduce the use of hazardous chemicals overall.

Lastly, the science of chemistry has been continuing to increase its capability to manipulate molecules and molecular transformation. As we have learned more about the hazards of chemicals, chemistry has developed powerful structure/activity relationship (SAR) tools that link hazard and molecular construction. With this knowledge chemists are now able to do what they do best, which is to manipulate molecules, and in this case use that manipulation to reduce hazard. Therefore, from a scientific point of view the answer to "why now for Green Chemistry" is because now we can. We now have scientific tools to address our environmental concerns at the molecular level.

Green Chemistry and Solvents

One accepted definition of a solvent is "Any substance that dissolves another substance so that the resulting mixture is a homogeneous solution. Solvents are some of the most ubiquitous classes of chemicals throughout society and the chemical industry. This is because the applications of chemicals in this class are so broad and varied. Solvents are used in manufacturing as a reaction medium for chemical transformations and also in a wide variety of processing steps in many industrial sectors. Solvents have been used for thousands of years in separations and extraction procedures as well as cleaning solutions. There are products as varied as paint to children's toys that use solvents in the product formulation.

When selecting a solvent for one of these widely varied applications, one needs to choose it based on the properties required for a particular application. These properties can be:

- Solubility
- Polarity
- Viscosity
- Volatility
- Other

However, just as one needs to select from these properties in order to meet certain performance criteria, Green Chemistry would suggest that reduced hazard is equally a performance criterion that needs to be met in the selection of a solvent. Some of the types of hazard that are common with a wide range of widely used solvents include the following:

- Inherent toxicity
- Flammability
- Explosivity
- Stratospheric ozone depletion
- Atmospheric ozone production
- Global warming potential

Just as one needs to optimize the classical physico-chemical property balance to select the proper solvent for a specified application, one also needs to optimize the selection or design of a new solvent or solvent system such that the performance criterion of reduced hazard is also met.

Traditional Solvents

Because of the wide range of hazards that can be exhibited by the large class of chemicals known as solvents, they have also tended to be some of the most highly regulated substances in the chemical industry. In the United States alone, the following laws/regulations/protocols have impacted and controlled the manufacture, use or disposal of solvents:

- Clean Air Act Amendments
- Clean Water Act
- Toxic Release Inventory
- Toxic Substances Control Act
- Montreal Protocol

When discussing the types of solvents that are regulated for various hazards we should review the common traditional solvents used so widely in products and processes. Some of the historically larger solvents would include the following:

- Chloroform, carbon tetrachloride, methylene chloride, perchloroethylene
- Benzene, toluene, xylene
- Acetone, ethylene glycol, methyl ethyl ketone
- CFCs

In contrast to these traditional solvents, Green Chemistry is using a variety of techniques to design new solvents, solvent systems and new ways of using known solvents to reduce or eliminate the intrinsic hazard associated with the tradition solvents and solvent systems.

Green Chemistry Alternatives

Using the principles of Green Chemistry, researchers in industry and academia are developing new solvents or solvent systems which reduce the intrinsic hazards associated with traditional solvents. In some cases new substances are being designed and developed to be used as solvents while in other cases some of the best known and characterized substances in the world are finding new applications as solvents. Of course, using no solvent at all in certain circumstances can be the ultimate solution to minimizing solvent associated hazards.

Some of the leading areas of work in alternative green solvents include the following:
- Aqueous solvents
- Supercritical or dense phase fluids
- Ionic liquids
- Immobilized solvents
- Solventless conditions
- Reduced Hazard Organic solvents
- Fluorous Solvents

Aqueous Solvents

Of course there is no better known substance nor a more widely used solvent in the world than water. It is also of course, one of the most innocuous substances on the planet. While water is also one of the most widely used solvents certainly by nature and also by man, there has been a trend in the last century to develop industrial processes where water was not considered an adequate or in some cases, even possible solvent. Many of these applications involved the chemical industry and the production of specialty chemicals.

What is now being discovered is that even in applications that for many decades were outside the range in which water could be used as a solvent, new processes are now being developed employing aqueous based systems. These systems use water effectively, efficiently, and with marked reduction in hazard to human health and the environment.

Supercritical or Dense Phase Fluids

An extremely important area of green solvent research and development is the area of supercritical or dense phase fluids. In this case, innocuous, extremely well characterized substances such as carbon dioxide are used as the solvent system. By changing the conditions to either approaching or past the critical point, a unique set of properties are attained that can be used for solvent purposes. This way, an innocuous substance such as carbon dioxide, or even supercritical water, can be used as a solvent with reduced hazard in a wide range of applications ranging from reaction medium to cleaning solvent.

The work in this volume ranging from catalytic systems to polymerization reactions shows the breadth of applicability of dense phase fluids as clean solvents.

Ionic Liquids

Ionic liquids, or room temperature molten salts have thus far taken a somewhat different approach to reducing potential hazard to human health and the environment. The concerns for many of the traditional solvents that we have mentioned are able to manifest their hazard as a function of their volatility. For example, CFCs are able to damage the ozone layer because they are capable of reaching the ozone layer and carrying out the necessary reactions. Other solvents that manifest toxicity through human inhalation are able to do this because they are volatile enough to have a concentration in the air sufficient for humans to respire. Ionic liquids have been designed such that the vapor pressure is so negligible that many of the researchers characterize it as "having no vapor pressure". Since this is the case, it is not able to manifest an intrinsic hazard that it might possess, thereby reducing the potential for harm to human health and the environment. The next step in the evolution of ionic liquids that is currently being vigorously pursued is the design of the molecular structures such that the inherent toxicity is also lowered as well as the potential for the ionic liquid to manifest the hazard. In addition, the environmental, economic and logistical challenges of isolating the products of a reaction or process will be met through development of Green Chemistry techniques in ionic liquids.

The IL research reported in this volume illustrates the challenges both scientific and commercial as well as the potential for radically changing industrial chemical processes through the use of ionic liquids.

Immobilized Solvents

Immobilized solvents or solvent molecules tethered to a polymeric backbone follow the same logic as the ionic liquids. By creating a system where a known solvent, e.g., THF, is tethered properly, it can still maintain its solvency but is incapable of manifesting any hazard by exposing humans or the environment.

While these types of solvents may initially appear to be expensive and difficult to handle, research is being done in specialty areas of fine chemicals and pharmaceuticals to identify possible niche applications.

Solventless Systems

The complete elimination of a solvent can obviously be the ultimate in reducing or eliminating the hazards associated with solvent usage. Much work is ongoing evaluating systems where the reagents for a particular process are also serving as the solvent, and many coating systems are now being designed to be solventless as well. However, this is an area that, just as the other solvent alternatives, there must be consideration of so-called life cycle effects. While solvent is removed in one step to reduce that hazard, it mustbe determined that the hazard is not actually increased elsewhere by needing more rigorous and material intensive separations down stream. This is a general guideline in the development of any Green Chemistry alternative that applies not only to solventless systems but also to all innovations and applications.

Reduced Hazard Organic Solvents

Organic solvents of various types do not all pose the same type or degree of hazard and toxicity. The investigation and application of various next generation organic solvents is being pursued. The use of various esters, e.g. Dibasic esters (DBE) are being pursued for their application in various types of chemistries and in industrial processes. The benefits of reduced toxicity and yet the similarity to some existing and popular solvent systems make these an extremely attractive area of investigation and implementation.

Fluorous Solvents

Since the time of the seminal work in fluorous solvents by Horvath (7), there have been extensive investigations of the applications of these new systems to carry out transformations in a wide range of systems. The application of these systems has advantages for certain types of reduced toxicity while the environmental impacts of their persistence continue to be evaluated as a class. Appropriate application of these solvents systems as illustrated in this book, and the innovations on the scope of their utilization, e.g., biological systems, continues at an aggressive pace.

Conclusion

The use of Green Chemistry to design new solvents and new solvent/solventless systems that are more environmentally benign and more economically beneficial is flourishing. There are many exciting breakthroughs in both research and industrial application in this field of endeavor and they are portrayed in this volume. New solvent systems and solventless processes are being developed to make one of the most widespread categories of chemicals, solvents, more environmentally benign.

References

1. Ritter, S., Green Chemistry, *Chemical and Engineering News*, **79** (29), 27-34, July 16, 2001.
2. Ritter, S., Accepting the Green Chemistry Challenge, *Chemical and Engineering News*, **79** (27), 24-27, July 2, 2001.
3. Anastas, P.T. and Warner, J.C., **Green Chemistry Theory and Practice**, Oxford University Press, 1998.
4. Collins, T.J., Green Chemistry, **MacMillens Encyclopedia of Chemistry**, McMillen, New York, 1997.
5. Anastas, P.T., Heine, L.G., and Williamson, T.C., **Green Engineering**, American Chemical Society, Symposium Series 766, 1-8, 2001.
6. Anastas, P.T., and Lankey, R., Green Chemistry and Life Cycle Analysis, *Journal of Green Chemistry*, **289**, (2), 2000.
7. Horvath, I.T., Rabai, J. *Science* **266**: (5182) 72-75 OCT 7 1994

Chapter 2

Ionic Liquids: Green Solvents for the Future

Martyn J. Earle and Kenneth R. Seddon

The QUILL Centre, Stranmillis Road, The Queen's University of Belfast, Northern Ireland BT9 5AG, United Kingdom

Ionic liquids can be used for clean synthesis in a variety of reactions. Here we present a review of the type of chemical reactions that can be carried out in ionic liquids. These include commercially important reactions carried out in chloroaluminate(III) ionic liquids; the isomerisation of fatty acids and the cracking of polyethylene to low molecular weight alkanes. Also, a growing number of reaction types can be performed in neutral ionic liquids. These include the Heck reactions, Diels-Alder reaction, nucleophilic displacement reactions (S_N2 reaction), and Friedel-Crafts reactions.

Introduction

The vast majority of chemical reactions have been performed in molecular solvents such as water or hydrocarbons. Much of our understanding of chemistry is based upon the behaviour and interaction of molecules in these solvents. Recently, a new class of solvent has emerged that consists entirely of ions. They were initially referred to as "room temperature molten salts", but the name "ionic liquid" has now become almost universal. As the name suggests, these materials are liquid at or near to room temperature. They are made up of two components; the anion and the cation. Where there is only one type of anion and one type of cation present, the ionic liquid is a simple salt. As both the anion and cation can be varied, these solvents can be designed for a particular end use in mind or for a particular set of properties. This has lead to the term "Designer Solvents" being applied to ionic liquids (1). To many chemists, performing reactions in ionic liquids may seem daunting. Compounded with this is the fact that the range of ionic liquids or potential ionic liquids available is very large. However, many scientists have found that performing reactions in ionic liquids is not at all difficult and often turns out considerably more practical than performing similar reactions in conventional organic solvents. This is particularly the case when considering reactions normally carried out in dipolar aprotic solvents like DMSO. It turns out that these noxious solvents can often be directly replaced with an ionic liquid.

A brief glance at the reference section of this review is testimony to the rapid growth of organic chemistry in ionic liquids, in that 60% of the references have only been published in the last two years. This rapidly expanding field of research has been reviewed by a number of authors, including Welton (2), Holbrey (3), Earle (4), Rooney (5) and Seddon (6). The earliest report of a room-temperature ionic liquid in the chemical literature was in 1914 when [EtNH₃][NO₃] (m.p. 12 °C) (7) was discovered. Interest did not develop until the discovery of binary ionic liquids made from mixtures of aluminium(III) chloride and initially N-alkylpyridinium chlorides (8). This was followed in 1982 with the introduction of 1,3-dialkylimidazolium chlorides (9); the derivatised imidazolium cation is perhaps the best cation for forming ionic liquids. In general, ionic liquids consist of a salt where one or both the ions are large, and have a low degree of symmetry (6). These factors reduce the lattice energy of the crystalline form of the salt and hence lower the melting point (10).

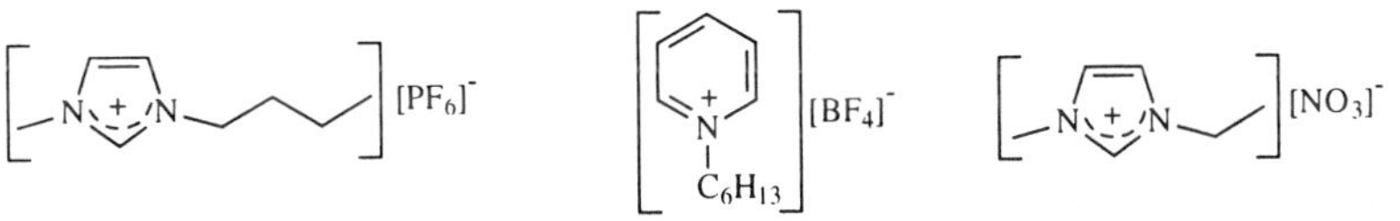

Figure 1. Examples of simple room temperature ionic liquids.

Ionic liquids fall into two categories, which are simple salts (made of a single anion and single cation) and binary ionic liquids (salts where an equilibrium is involved). For example, [EtNH₃][NO₃] is a simple salt whereas mixtures of aluminium(III) chloride and 1,3-dialkylimidazolium chlorides (a binary ionic liquid system) contain several different ionic species. The melting point and properties depend upon the mole fractions of the aluminium(III) chloride and 1,3-dialkylimidazolium chlorides present. Examples of ionic liquids consisting of a simple salt are given in Figure 1.

Freemantle (1) has described ionic liquids as "designer solvents", whose properties can be adjusted to suit the requirements of a particular process. By making changes to the structure of either the anion, or the cation, or both, properties such as solubility, density, refractive index and viscosity can be adjusted to suit requirements. Two examples of this are given in Figure 2, where the melting points of 1-alkyl-3-methylimidazolium tetrafluoroborates are plotted against alkyl chain length (*11*). Also, the melting points of 1-alkyl-3-methylimidazolium hexafluorophosphates (*12*) give rise to a similar type of behaviour. It has been observed that a liquid crystalline phase can be formed for both these systems. This occurs where the alkyl chain length is greater than twelve carbon atoms. Another important property that changes with structure is the miscibility of these ionic liquids with water. For example, 1-alkyl-3-methylimidazolium tetrafluoroborate salts are miscible with water at 25 °C where the alkyl chain is less than 6, but at or above 6 carbon atoms, they form a separate phase when mixed with water. This behaviour can be of substantial benefit when carrying out solvent extractions or product separations, as the relative solubilities of the ionic and extraction phase can be adjusted to assist with the separation (*13*). Also, separation of the products can be achieved by other means such as distillation (usually under vacuum) or by steam distillation.

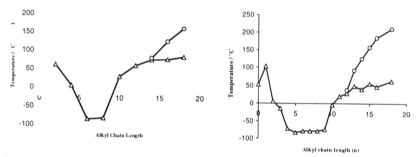

Figure 2. The variation of melting point with alkyl chain length for 1-alkyl-3-methyl imidazolium hexafluorophosphates (left) and tetrafluoroborates (right). The triangle represents the solid to liquid or liquid crystal transition temperature and the circle represents the melting point of the liquid crystal.

Reactions In Chloroaluminate(III) Ionic Liquids

The chemical behaviour of Franklin acidic chloroaluminate(III) ionic liquids (where $X(AlCl_3) > 0.50$) (*11*) is that of a powerful Lewis acid. As might be expected, it catalyses reactions that are conventionally catalysed by aluminium(III) chloride, without suffering the disadvantage of the low solubility of aluminium(III) chloride in many solvents. Indeed, chloroaluminate(III) ionic liquids are exceptionally powerful solvents (*4*). These ionic liquids are straightforward to prepare and are made by mixing the appropriate organic halide salt with aluminium(III) chloride (*14*). This results in the two solids melting together (exothermically) to form the ionic liquid. It should be noted that these chloroaluminate(III) ionic liquids are very moisture sensitive and their synthesis and use must be performed in an inert atmosphere.

A classical reaction catalysed or promoted by Lewis acids is the Friedel-Crafts reaction, which was found to work efficiently in chloroaluminate(III) ionic liquids (15). A number of commercially important fragrance molecules have been synthesised by Friedel-Crafts acylation reactions in these ionic liquids (16). Traseolide® (5-acetyl-1,1,2,6-tetramethyl-3-isopropylindane) (Figure 3) and Tonalid® (6-acetyl-1,1,2,4,4,7-hexamethyltetralin) have been made in high yield in the ionic liquid [emim]Cl-AlCl$_3$ (X = 0.67). The acylation of naphthalene, the ionic liquid gives the highest known selectivity for the 1-position (16). The acetylation of anthracene at 0 °C was found to be a reversible reaction. The initial product of the reaction of acetyl chloride (1.1 equivalents) with anthracene is 9-acetylanthracene, formed in 70 % yield in under 5 minutes.

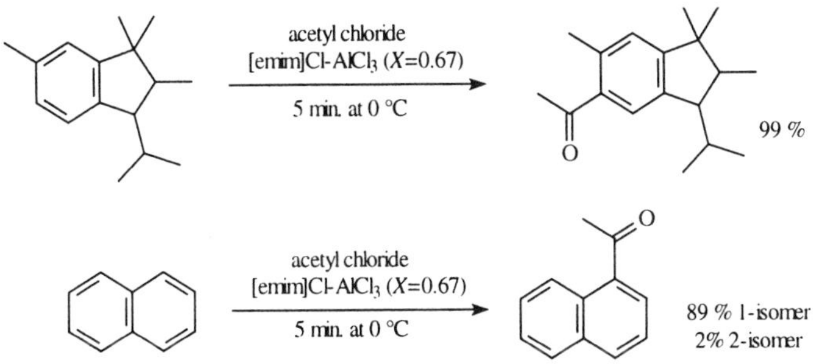

Figure 3. The acetylation of 1,1,2,6-tetramethyl-3-isopropylindane and naphthalene in [emim]Cl-AlCl$_3$ (X = 0.67).

The 9-acetylanthracene was then found to undergo a diacetylation reaction, giving the 1,5- and 1,8-diacetylanthracene and anthracene after 24 hours. The reaction of 9-acetylanthracene with pure [emim]Cl-AlCl$_3$ (X = 0.67) did not occur, until a proton source such as water or HCl was added. It then gave a mixture of anthracene (48 %) and 1,5- and 1,8- diacetylanthracene (47 %). The variation of composition with time in the reaction of anthracene with 2.1 equivalents acetyl chloride is given in Figure 4. As can be seen, the intermediate 1- and 9-acetylanthracene compounds are formed and consumed during the reaction.

Cracking and isomerisation reactions occur readily in acidic chloroaluminate(III) ionic liquids. A remarkable example of this is the reaction of polyethylene, which is converted to a mixture of gaseous alkanes with the formula (C_nH_{2n+2}, where n = 3-5) and cyclic alkanes with a hydrogen to carbon ratio of less than two (17) (Figure 5).

14

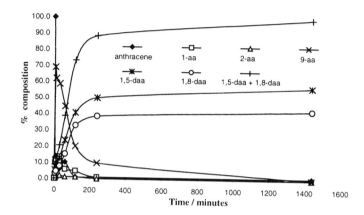

Figure 4. The variation of composition with time for the reaction of 2.1 equivalents of acetyl chloride with anthracene in [emim]Cl-AlCl₃ (X = 0.67). Note: aa = acetylanthracene and daa = diacetylanthracene.

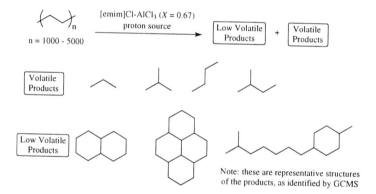

Figure 5. Isomerisation and cracking reactions of polyethylene in chloroaluminate(III) ionic liquids.

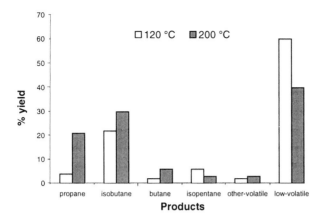

Figure 6. The products from the ionic liquid cracking of high-density polyethylene (n = 5000) at 120 °C and 200 °C.

The distribution of the products obtained from this reaction depends upon the reaction temperature (Figure 6) and differs from other polyethylene recycling reactions in that aromatics and alkenes are not formed in significant concentrations. Another significant difference is that this ionic liquid reaction occurs at temperatures as low as 90 °C, whereas conventional catalytic reactions require much higher temperatures, typically 300-1000 °C (*18*). A similar cracking reaction has been reported by SECR Defense for lower molecular weight hydrocarbons in chloroaluminate(III) ionic liquids (*19*).

A similar reaction occurs with fatty acids (such as stearic acid) or methyl stearate, which undergo isomerisation, cracking, dimerisation, and oligomerisation reactions. This has been used to convert solid stearic acid into the more valuable liquid isostearic acid (*20*). The isomerisation and dimerisation of oleic acid and methyl oleate has also been found to occur in chloroaluminate(III) ionic liquids (*21*). These reactions are shown in Figure 7, where the structures given are representative of the products obtained as a large number of isomers of monomer, dimer and trimer are obtained. In the case of the trimer, typically two, three or four carbocyclic rings are observed.

The dimerisation and oligomerisation of olefins in the presence of homogeneous nickel(II) catalysts has been studied extensively in chloroaluminate(III) and alkylchloroaluminate(III) ionic liquids (*22,23*). Few catalysts are known that catalyse the linear dimerisation and oligomerisation of C_4-olefins. Linear C_8-olefin dimers are highly desirable intermediates for the production of C_9-plasticizers, exhibiting better thermal properties than those produced from highly branched C_8-olefin dimer feedstock. This has been developed commercially in the IFP Difasol process (*1*). The products of these reactions are insoluble in the ionic liquid and can be separated by decantation, leaving the ionic liquid and catalyst behind and hence the catalyst and solvent can be reused.

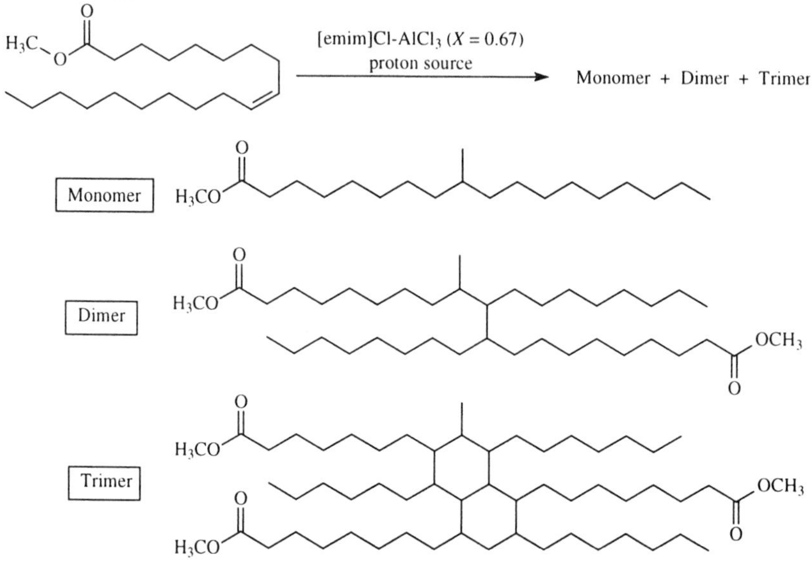

Figure 7. The isomerisation of methyl oleate in [emim]Cl-AlCl₃ (X = 0.67).

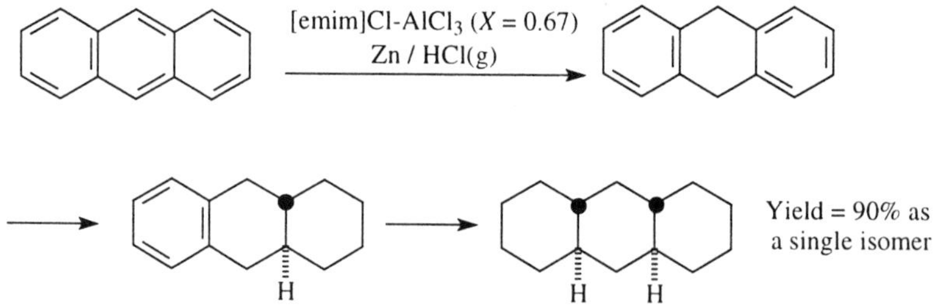

Figure 8. The sequence of reduction of anthracene to perhydroanthracene.

Polycyclic aromatic hydrocarbons dissolve in chloroaluminate(III) ionic liquids to form highly coloured paramagnetic solutions (24). The addition of a reducing agent such as an electropositive metal and a proton source results in the selective hydrogenation of the aromatic compound. For example pyrene and anthracene can be reduced to perhydropyrene and perhydroanthracene at ambient temperatures and pressures. Interestingly, only the thermodynamically most stable isomer of the product is obtained (25). This contrasts with catalytic hydrogenation reactions, which require high temperatures and pressures and an expensive platinum oxide catalyst and give rise to an isomeric mixture of products (26). By careful monitoring of the reduction in the ionic liquid, a number of intermediate products can be isolated and the sequence of the chemical reduction process can be determined (Figure 8).

Chloroaluminate(III) ionic liquids have recently been used for a number of other reaction types. One example is in the acylative cleavage of ethers. Singer *et al.* have shown that benzoyl chloride reacts with diethyl ether to give ethyl benzoate (27). Another common reaction is the chlorination of alkenes to give dihaloalkanes. Patell *et al.* has reported that the addition of chlorine to ethene in acidic chloroaluminate(III) ionic liquids gave 1,2-dichloroethane (28). Other recently reported reactions include hydrogenation reactions of cyclohexene using Wilkinson's catalyst, [RhCl(PPh₃)₃], in basic chloroaluminate(III) ionic liquids (29) but neutral ionic liquids are preferred for this type of reaction, due to their ease of handling and lower moisture sensitivity. Also, esterification (30) and aromatic alkylation reactions (31) have appeared in the patent literature.

Neutral Ionic Liquids

Chloroaluminate(III) ionic liquids are excellent catalysts and solvents in many processes, but suffer from several disadvantages, such as their moisture sensitivity and the difficulty of separation of products containing heteroatoms from the ionic liquid, whilst leaving the ionic liquid intact. Hence research is shifting to the investigation of ionic liquids that are stable to water. This allows for straightforward product separation and ease of handling. In order to develop chemistry in ionic liquids, and increase the robustness of processes, water-stable ionic liquids have become increasingly important. In particular, a number of ionic liquids have been found to be hydrophobic (immiscible with water), but readily dissolve many organic molecules (with the exception of alkanes, ethers and alkylated aromatic compounds such as toluene). An example of this is the ionic liquid [bmim][PF₆] (14,32), ([bmim] = 1-butyl-3-methylimidazolium) which forms triphasic solutions with alkanes and water (33).

This multiphasic behaviour has important implications for clean synthesis and is analogous to the use of fluorous phases in some chemical processes (34). For example, transition metal catalysts can be exclusively dissolved in the ionic liquid. This allows products and by-products to be separated from the ionic liquid by solvent extraction with either water or an organic solvent. This is of key importance when using precious metal catalysts or catalysts with expensive ligands, enabling both the ionic liquid and catalyst to be recycled and reused. Alternatively, some volatile products can be separated from the ionic liquid and catalyst by distillation. This is because the ionic liquid has effectively no vapour pressure and therefore cannot be lost. There also exists the possibility of

18

extraction with supercritical solvents. A recent example is the use of supercritical carbon dioxide to extract naphthalene from [bmim][PF₆] (35).

One of the key findings when carrying out reactions in neutral ionic liquids is that special conditions are not usually required. There is often no need to exclude water from the reaction vessel, or to carry the reaction out in an inert atmosphere such as under dinitrogen or argon. This, combined with the ability to design the ionic liquid to allow for easy separation of the product, makes reactions in ionic liquids extremely straightforward to perform.

Neutral ionic liquids have also been found to be excellent solvents for the Diels-Alder reaction (36,37) giving significant rate enhancement over molecular solvents, including water, which is normally considered to enhance the rate of this chemical reaction (38). The conversion with time of ethyl acrylate and cyclopentadiene is given in Figure 9. As can be seen, the reactions in ionic liquids are marginally faster than in water and are considerably faster than in the organic solvent diethyl ether. The selectivities of the various isomers in this reaction can be improved from 4 : 1 to 20 : 1 by the addition of a mild Lewis acid such as zinc(II) iodide to the ionic liquid (Table 1). It is clear that the Lewis acid allows the reaction to be carried out at room temperature with shorter reaction times and with better yields and selectivities. One of the key benefits of this if that the ionic liquid and catalyst can be recycled and reused after solvent extraction or direct distillation of the product from the ionic liquid. The reaction has also been carried out in chloroaluminate(III) ionic liquids, but the moisture sensitivity of these systems is a major disadvantage (39).

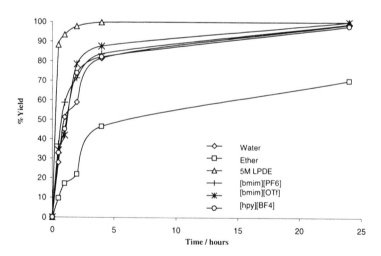

Figure 9. The reaction ethyl acrylate (10 mmol) and cyclopentadiene (15 mmol) in 2.5 g of ether, water, 5M lithium perchlorate in diethyl ether (LPDE), [bmim][PF₆], [bmim][OTf] or [hpy][BF₄] ([hpy] = 1-hexylpyridinium).

Table 1. The reactions of isoprene with various dienophiles in ionic liquids.

Entry	Solvent	Dienophile	Product	Temp /°C	Time / h.	% Yield	4- to 3- Isomer ratio
1	[bmim][PF$_6$]	DMAD	1	20	15	45	
2	[bmim][PF$_6$]	DMAD	1	80	2	98	
3	[bmim][OTf]	ethyl acrylate	2	70	24	97	2.5 : 1
4	[bmim][OTf]	ethyl acrylate	2	20	24	16	4.5 : 1
5	[bmim [PF$_6$]	3-butene-2-one	3	20	18	11	4.0 : 1
6	[bmim][PF$_6$] [a]	3-butene-2-one	3	20	6	98	20.0 : 1
7	[bmim][BF$_4$] [b]	ethyl acrylate	2	20	2	90	40 : 1
8	[bmim][BF$_4$]	ethyl acrylate	2	20	24	14	4.4 : 1

[a] 5 mol % ZnI$_2$ added

[b] 10 mol % BF$_3$·OEt$_2$ added.

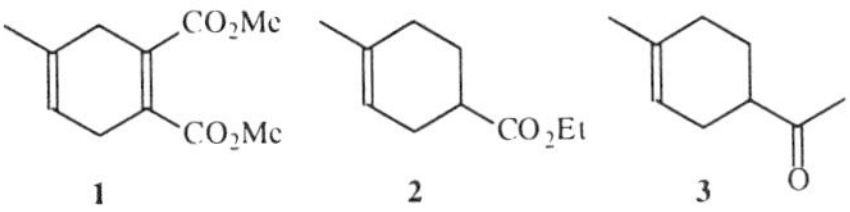

This Lewis acid in ionic liquid methodology has also been used for Friedel-Crafts alkylation reactions. Song (40) has reported that scandium(III) triflate in [bmim][PF$_6$] acts as an alkylation catalyst in the reaction of benzene with hex-1-ene. Also Raston has reported an acid catalysed Friedel-Crafts reaction (41) where 3,4-dimethoxyphenylmethanol has been cyclised to cyclotriveratrylene.

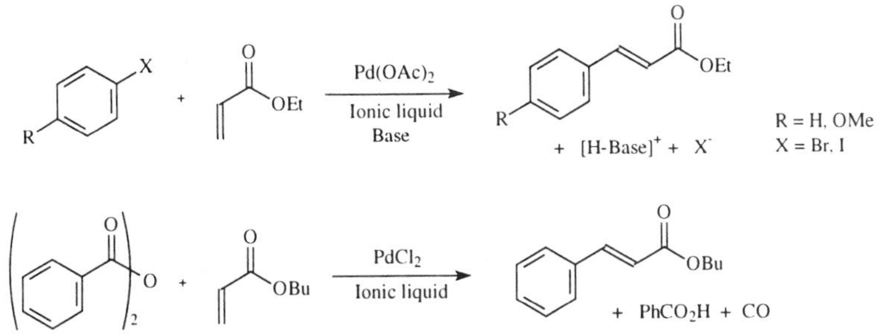

Figure 10. The Heck reactions of aryl halides and anhydrides in ionic liquids.

A particularly useful reaction is the palladium-catalysed coupling of aryl halides with alkenes (the Heck reaction), and was first reported in molten alkylammonium and phosphonium salts (42). It has been found that many palladium complexes dissolve in ionic liquids (43) and allow the products and by-products of the Heck reaction to be extracted with either water or alkane solvents (33). Also, separation of the products by distillation directly from the ionic liquid / catalyst is feasible where the products and by-products are volatile. This allows the expensive catalyst to be easily recycled as it remains in the ionic phase. This differs from conventional Heck reactions, in which the catalyst is usually lost at the end of the reaction, and also noxious dipolar aprotic solvents such as DMF are required. An alternate Heck reaction in ionic liquids uses aromatic anhydrides as a source of the aryl group (Figure 10). This has the advantage that the by-product of the reaction is the arylcarboxylic acid (which can readily be converted back to the anhydride) and that halide-containing waste is not formed.

Two types of ionic liquid / catalyst combinations were found to be effective, the first being based on chloride salts. A combination of N-hexylpyridinium chloride and palladium(II) acetate was found to be one of the most effective combinations (Note: 2H-imidazolium cationic salts do not work well). Here, the palladium(II) acetate interacts with the ionic liquid to form [hpy]$_2$[PdCl$_4$] (44), which is thought to be the catalytic species. The second system was a 1-alkyl-3-methylimidazolium hexafluorophosphate(V) or tetrafluoroborate(III) ionic liquid combined with palladium(II) acetate / Group 15 ligand catalyst. The ligand is needed to prevent the formation of imidazolylidine complexes with the palladium catalyst. The structures of these palladium carbene complexes have been reported by several authors (45,46) and have been found to be catalytically active in the Heck reaction (46). Similar findings were subsequently reported by Howarth in the reaction of 4-bromoanisole with alkyl acrylates, and confirmed that [bmim][PF$_6$] could be used as a replacement for DMF (47).

Other reactions that have been found to proceed under similar conditions, include palladium catalysed allylation and amination reactions in [bmim][BF$_4$] (48). An example is the alkylation of dimethyl malonate by 3-acetoxy-1,3-diphenylprop-1-ene. This was further developed by Toma et al. who used a chiral palladium catalyst (49) to obtain the product (Figure 11) in moderate enantioselectivities. Also, Welton has reported the Suzuki Coupling of aryl halides with aryl boronates (50).

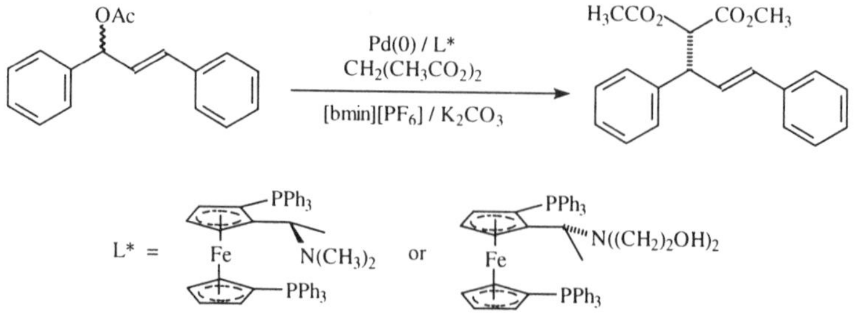

Figure 11. An asymmetric allylic substitution reaction in an ionic liquid.

One of the most common reactions in organic synthesis is the nucleophilic displacement reaction. Indole and 2-naphthol (*51*) undergo alkylation on the nitrogen and oxygen atoms respectively (Figure 12), when treated with an alkyl halide and base (usually NaOH or KOH) in [bmim][PF₆] (*52*). These reactions occur with similar rates to those carried out in dipolar aprotic solvents such as DMF or DMSO. The advantage of the ionic liquid process is that the products of the reaction can be extracted into an organic solvent such as toluene, leaving the ionic liquid behind. The by-product (sodium or potassium halide) of the reaction can be extracted with water and the ionic liquid recycled. This contrasts with the use of dipolar aprotic solvents, where they are difficult to remove from the product.

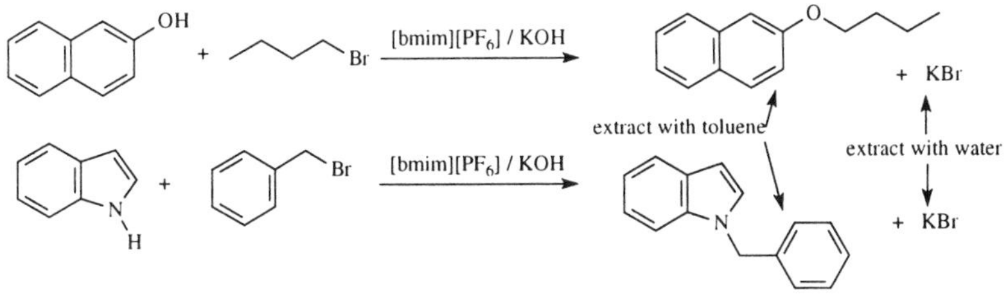

Figure 12. Alkylation reactions in [bmim][PF₆].

The addition of organometallic reagents to carbonyl compounds is an important reaction in organic chemistry, with the Grignard reaction being one example of this. Gordon and McClusky (*53*) have reported the formation of homoallylic alcohols, from the addition of allyl stannanes to aldehydes in the ionic liquids [bmim][BF₄] and [bmim][PF₆] (Figure 13). It was found that the ionic liquid could be recycled and reused over several reaction cycles.

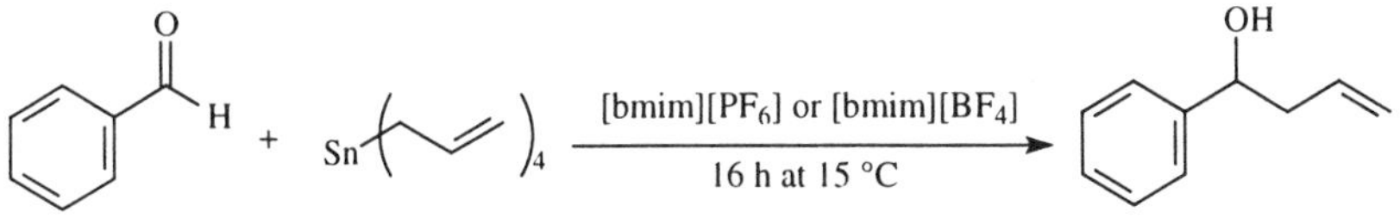

Figure 13. Allylation of aldehydes in [bmim][PF₆] or [bmim][BF₄].

Neutral ionic liquids have been extensively studied as solvents for hydrogenation reactions. A key advantage of ionic liquids is that homogeneous transition metal catalysts can be used and the products of the reaction can be easily separated from the ionic liquid and catalyst (*2*). Examples of this include the hydrogenation of cyclohexene (*54*), and the complete hydrogenation of benzene rings (*55*). Recently, asymmetric hydrogenation reactions have appeared. An example is in the synthesis of (S)-Naproxen (Figure 14) in the ionic liquid [bmim][BF₄] (*56*).

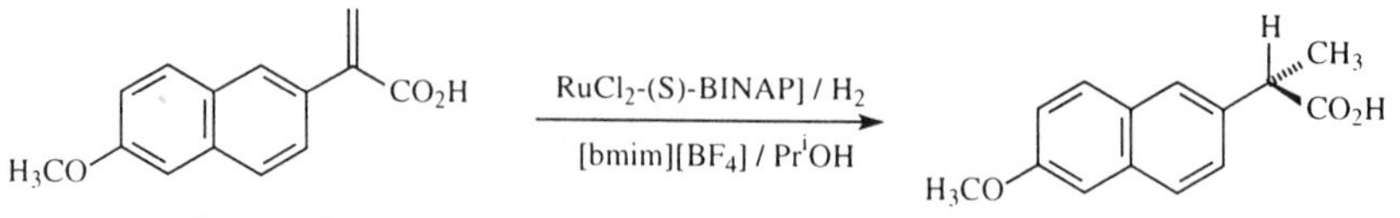

Figure 14. The synthesis of (S)-Naproxen in [bmim][BF₄].

A surprising use of ionic liquids is in the field of bio-transformations. This was first investigated Magnuson *et. al.* (*57*). Room temperature ionic liquids, such as [bmim][PF₆], have been used as direct replacements for conventional organic solvents in multiphase bioprocess operations. Examples include the liquid-liquid extraction of the antibiotic erythromycin and two-phase bio-transformation processes (*58*).

Pharmaceutical synthesis In Ionic Liquids

As a demonstration of the complete synthesis of a pharmaceutical in an ionic liquid, Pravadoline was selected as it combines a Friedel-Crafts reaction and a nucleophilic displacement reaction (Figure 15) (*59*). The alkylation of 2-methylindole with 1-(N-morpholino)-2-chloroethane occurs readily in [bmim][PF₆] at room temperature in 95% yield using potassium hydroxide as the base. The yield can be improved to 99% using the ionic liquid 1-butyl-2,3-dimethylimidazolium hexafluorophosphate ([bdmim][PF₆]).

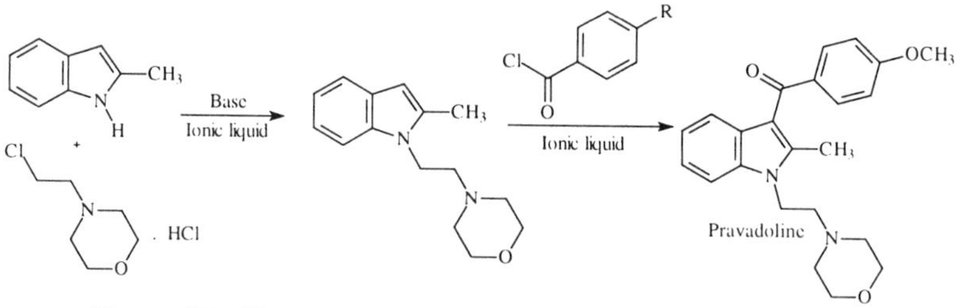

Figure 15. The complete synthesis of Pravadoline in [bmim][PF₆].

The Friedel-Crafts acylation of the product from the nucleophilic displacement reaction can be performed in chloroaluminate(III) ionic liquids at 0 °C, and was found to give the desired product in 70% yield. However, a large excess of the ionic liquid is needed and is lost at the end of the reaction. The best results occur in [bmim][PF₆] at 150 °C with no catalyst. This considerably reduces the amount of waste formed. When KOH is used as the base in both steps of the reaction, the only waste from the whole synthesis is aqueous potassium chloride, which compares favourably with the original synthesis that produces large amounts of acidic aluminium waste and use dipolar aprotic solvent (*60*). Other benefits are that the ionic liquid process does not require strictly anhydrous conditions or an inert atmosphere to carry the reaction out in.

Conclusions

We have shown that by choosing the correct ionic liquid, high product yields can be obtained and a reduced the amount of waste is produced in a given reaction. Often the ionic liquid can be recycled and this leads to a reduction the costs of the processes. It must be emphasised that reactions in ionic liquids are not difficult to perform and usually require no special apparatus or methodologies. The reactions are often quicker and easier to carry out than in conventional organic solvents. This leaves the whole of synthetic organic chemistry open for reinvestigation in ionic liquids, with many anticipated green benefits.

Acknowledgements

We are would like to thank the following for financial support: The QUILL Center (M.J.E.), the EPSRC and the Royal Academy of Engineering for the award of a Clean Technology Fellowship (K.R.S.).

References

1. Freemantle, M. *Chem. Eng. News* **1998 (30th March)**, *76*, 32-37.
2. Welton, T. *Chem. Rev.* **1999**, *99*, 2071-2083.
3. Holbrey, J.; Seddon, K. R. *Clean Prod. Proc.* **1999**, *1*, 223-236.
4. Earle, M. J.; Seddon, K. R. *Pure Appl. Chem.* **2000**, *72*, 1391-1398.
5. Rooney, D. W.; Seddon, K. R. In *The Handbook of Solvents;* Editor Wypych G.; ChemTech Publishing, New York, **2001**, pp 1459-1484.
6. Seddon, K. R. *J. Chem. Tech. Biotech.* **1997**, *68*, 351-356.
7. Walden, P. *Bull. Acad. Imper. Sci. (St. Petersburg)* **1914**, 1800.
8. Hurley, F. H.; Weir, T. P.; *J. Electrochem. Soc.* **1951**, *98*, 203-204.
9. Wilkes, J. S.; Leviskey, J. A.; Wilson, R. A.; Hussey, C. L. *Inorg. Chem.* **1982**, *21*, 1236-1264.
10. Seddon, K. R. In *Molten Salt Forum: Proceedings of 5th International Conference on Molten Salt Chemistry and Technology*; Editor Wendt, H. **1998**, Vol. 5-6, pp 53-62.
11. Holbrey, J. D.; Seddon, K. R. *J. Chem. Soc., Dalton Trans.* **1999**, 2133-2139.
12. Gordon, C. M.; Holbrey, J. D.; Kennedy, A. R.; Seddon, K. R. *J. Mater. Chem.* **19 98**, *8*, 2627.
13. Visser, A.E.; Swatloski R.P.; Rogers, R.D. *Green Chem.* **2000**, *2*, 1-4.

14. Chum, H. L.; Koch, V. R.; Miller, L. L.; Osteryoung, R. A. *J. Am. Chem. Soc.* **1975**, *97*, 3264-3267.
15. Boon, J. A.; Levisky, J. A.; Pflug, J. L.; Wilkes, J. S. *J. Org. Chem.* **1986**, *51*, 480-483.
16. Adams, C. J.; Earle, M. J.; Roberts, G.; Seddon, K. R. *Chem. Commun.* **1998**, 2097-2098.
17. Adams, C. J.; Earle, M. J.; Seddon, K. R. *Green Chem.* **2000**, *2*, 21-24.
18. Westerhout, R. W. J.; Waanders, J.; Kuipers, J. A. M.; Van Swaaij, W. P. *Ind. Eng. Chem. Res.* **1998**, *37*, 2293-2300.
19. Barnes, P. N.; Grant K. A.; Green K. J.; Lever N. D. J. *World Patent* WO 0040673, **2000**.
20. Adams, C. J.; Earle, M. J.; Hamill, J.; Lok, C. M.; Roberts, G.; Seddon, K. R *World Patent* WO 9807680, **1998**.
21. Adams, C. J.; Earle, M. J.; Hamill, J.; Lok, C. M.; Roberts, G.; Seddon, K. R. *World Patent* WO 9807679, **1998**.
22. Ellis, B.; Keim, W.; Wasserscheid, P. *Chem. Commun.* **1999**, 337-338.
23. Chauvin, Y.; Olivier, H.; Wyrvalski, C. N.; Simon, L. C.; de Souza, R. F. *J. Catal.* **1997**, 275-278.
24. P. Tarakeshwar, J. Y. Lee, K. S. Kim. *J. Phys. Chem. A* **1998**, *102*, 2253-2255.
25. Adams, C. J.; Earle, M. J.; Seddon, K. R. *Chem. Commun.* **1999**, 1043-1044.
26. Dalling, D. K.; Grant, D. M. *J. Am. Chem. Soc.* **1974,** *96*, 1827-1834.
27. Green, L.; Hemeon, I.; Singer, R. D. *Tetrahedron Lett.* **2000**, *41*, 1343-1345.
28. Patell, Y.; Winterton N.; Seddon K. R. *World Patent,* WO 0037400, **2000**.
29. Suarez, P. A. Z.; Dullius, J. E. L.; Einloft, S.; de Souza, R. F.; Dupont, J. *Polyhedron* **1996**, *15*, 2127-2129.
30. Ma, Z.; Deng, Y.; Shi, F. *Canadian Patent,* CN 1247856, **2000**.
31. Hodgson, P. K. G.; Morgan, M. L. M.; Ellis B.; Abdul-Sada, A. A. K.; Atkins. P.; Seddon K. R. *US Patent,* US 5994602, **1999**.
32. Huddleston, J. D.; Willauer, H. D.; Swatloski, R. P.; Visser, A. E.; Rogers, R. D. *Chem. Commun.* **1998**, 1765-1766.
33. Carmichael, A. J.; Earle, M. J.; Holbrey, J. D.; McCormac, P. B.; Seddon, K. R. *Org. Lett.* **1999**, *1*, 997-1000.
34. Barthel-Rosa, L. P.; Gladysz, J. A. *Coord. Chem. Rev.* **1999**, *192*, 587-605.
35. Blanchard, L. A.; Hancu, D.; Beckman, E. J.; Brennecke, J. F. *Nature* **1999**, *399*, 28-29.
36. Earle, M. J.; McCormac, P. B.; Seddon, K. R. *Green Chem.* **1999**, *1*, 23-25.
37. Fisher, T.; Sethi, A.; Welton, T.; Woolf, J. *Tetrahedron Lett.* **1999**, *40*, 793-795.

38. Rideout, D. C.; Breslow, R. J. *J. Am. Chem. Soc.* **1980**, *102*, 7816-7822.
39. Lee, C. W. *Tetrahedron Lett.* **1999**, *40*, 2461-2462.
40. Song, C. E.; Shim, W. H. Roh, E. J.; Choi, J. H. *Chem. Commun.* **2000**, 1695-1696.
41. Scott, J. L.; MacFarlan, D. R.; Raston, C. L.; Teoh, C. M. *Green Chem.* **2000**, *2*, 123-126
42. Kaufmann, D. E., Nouroozian M, Henze H. *Synlett* **1996**, *11*, 1091.
43. Herrmann, W. A.; Bohn V. P. W. *J. Organomet. Chem.* **1999**, *572*, 141-142.
44. Earle, M. J.; Nieuenhuyzen, M.; Seddon, K. R. *Unpublished results.*
45. McGuinness, D. S.; Cavell, K. J.; Skelton, B. W.; White, A. H. *Organometallics*, **1999**, *18*, 1596-1598.
46. Xu, L.; Weiping, C.; Xiao, J. *Organometallics*, **2000**, *19*, 1123-1125.
47. Howarth, J.; Dallas, A. *Molecules*, **2000**, *5*, 851-853.
48. Chen, W.; Xu, L.; Chatterton, C.; Xiao, J. *Chem. Commun.* **1999**, 1247-1248.
49. Toma, S.; Gotov, B.; Kmentová, I.; Solcániová, E. *Green Chem.* **2000**, *2*, 149-151.
50. Mathews, C. J.; Smith, P. J.; Welton, T. *Chem. Commun.*, **2000**, 1249-1250.
51. Badri M., Brunet J. J. *Tetrahedron Lett.* **1992**, *33*, 4435-4438.
52. Earle, M. J.; McCormac, P. B.; Seddon, K. R. *Chem. Commun.* **1998**, 2245-2246.
53. C. M. Gordon, A. McClusky. *Chem. Commun.* **1999**, 143-144.
54. Suarez, P. A. Z.; Dullius, J. E. L.; Einloft, S.; de Souza, R. F.; Dupont, J. *Inorg. Chim. Acta.* **1997**, *225*, 207-209.
55. Dyson, P. J.; Ellis, D. J.; Parker, D. G.; Welton, T. *Chem. Commun.* **1999**, 25-26.
56. Monteiro, A. L.; Zinn, F. K.; de Souza, R. F.; Dupont, J. *Tetrahedron Asymmetry* **1997**, *8*, 177-179.
57. Magnuson D. K., Bodley, J. W., Evans D. F. *J. Solution Chem.* **1984**, *13*, 583-587.
58. Cull, S. G.; Holbrey, J. D.; Vargas-Mora, V.; Seddon, K. R.; Lye, G. T. *Biotechnology and Bioengineering* **2000**, *69*, 227-232.
59. Earle, M. J.; McCormac, P. B.; Seddon, K. R. *Green Chem.* **2000**, *2*, 261-262.
60. Bell, M. R.; Dambra, T. E.; Kumar, V.; Eissenstat, M. A.; Herrmann, J. L.; Wetzel, J. R.; Rosi, D.; Philion, R. E.; Daum, S. J.; Hlasta, D. J.; Kullnig, R. K.; Ackerman, J. H.; Haubrich, D. R.; Luttinger, D. A.; Baizman, E. R.; Miller, M. S.; Ward, S. J. *J. Med. Chem.* **1991**, *34*, 1099-1110.

Chapter 3

Synthesis of Ionic Liquid and Silica Composites Doped with Dicyclohexyl-18-Crown-6 for Sequestration of Metal Ions

Rajendra D. Makote, Huimin Luo, and Sheng Dai[*]

Chemical Technology Division, Oak Ridge National Laboratory, P.O. Box 2008, Oak Ridge, TN 37831–6181

Abstract

A monolithic composite glass material that contains an ionic liquid, dicyclohexyl-18-crown-6, and sol-gel silica glass was synthesized and characterized . The selective extraction of Sr^{2+} by this solid-state sorbent from aqueous solutions was demonstrated.

Introduction

Macrocyclic crown ethers that form strong complexes with alkali and alkaline earth metal ions have been the subject of numerous investigations.[1, 2, 3] Various applications have been developed to make use of their selective complexation capabilities. These applications include ion-selective electrodes, spectrophotometrical determinations, and ion-selective liquid membranes. In recent years, there has been extensive interest in the development of selective solid-state sorbents for toxic metal ions based on these macrocyclic crown ethers.[4, 5, 6] Notably, sol-gel-derived membranes incorporating crown ether neutral carriers have been synthesized by Kimura, et al.[4] by means of sol-gel processing using tetraethoxysilane, diethoxydimethylsilane, and their corresponding alkoxysilylated neutral carriers. Zhang and Clearfield[5] have reported novel syntheses of crown ether-pillared α-zirconium phosphonate phosphates. Izatt, et al. have succeeded in covalently attaching a variety of crown ethers into a polymer matrix for chromatographic separation.[6]

Here, we report a methodology to immobilize crown ethers in a liquid nanophase that is entrapped in sol-gel silica for sequestration of metal ions. The nanophase is defined as any liquid domain with size in range of 1 nm to 50 nm. Here, the liquid nanophase is composed of a room-temperature ionic liquid. Ionic liquids are attracting increased attention worldwide because they promise significant environmental benefits.[7] Unlike the conventional solvents currently in use, ionic liquids are nonvolatile and therefore do not emit noxious vapors, which can contribute to air pollution and health problems for process workers. Accordingly, no losses are induced by vaporization when these ionic liquids are entrapped in solid matrices, even at high temperatures. Unique intrinsic properties of these liquids are that they consist *only of ions* and can be made *hydrophobic*! We have recently demonstrated that the novel dual properties of these new ionic liquids make them efficient solvents for the extraction of *ionic species* from aqueous solutions.[8] Whereas conventional solvent extraction of Sr^{2+} and Cs^+ using crown ethers and related extractants can deliver practical K_d value of up to two orders of magnitude (10^2), tests with ionic liquids as extraction solvents delivered values of K_d on the order of 10^4. These results clearly show the enormous potential that ionic liquids possess for increasing the extractive strength of ionophores such as crown ethers in fission-product separation applications. The same enhancement may hold also for ionic liquids entrapped in silica networks.

Ionic liquid and organic polymer composites have been recently synthesized.[9, 10] Specifically, Carlin and Fuller[9a] reported a novel application of a gas-permeable ionic-liquid polymer gel composed of 1-N-butyl-3-methylimidazolium hexafluorophosphate and poly(vinylidene fluoride)-hexafluoropropylene copolymer in catalytic heterogeneous hydrogenation reactions. Applications of these organic composite polymers in a variety of electrochemical devices have been demonstrated by Fuller et al.[9b] and Kosmulski et al.[10] Our interest lies in the synthesis of ionic-liquid and inorganic hybrid composite materials. These materials should be more stable under harsh environments because of the ceramic networks. The temperature stability is limited only by that of the ionic liquid.

Experimental

Dicyclohexyl-18-crown-6 (DC18C6), tetramethyl orthosilicate (TMOS), and formic acid (99%) were used as received from Aldrich. The ionic liquid used in this experiment is 1-ethyl-1-methyl imidazolium bis(trifluoromethyl) sulfonamide (EtMeIm$^+$Tf$_2$N$^-$), which was synthesized as reported previously and is known to be hydrophobic.[11] Acid-catalyzed sol-gel processes were utilized to synthesize the composite materials doped with crown ether.[12, 13] In a typical run, 50 mg of crown ether was dissolved into 1 mL of EtMeIm$^+$Tf$_2$N$^-$ ionic liquid. To this mixture was added 1 mL of TMOS and 2 mL of formic acid. The solution gelled after 1 day and gelation continued for 1 week. The volatile hydrolysis products (CH$_3$OH and HCOOCH$_3$) were vaporized. The final product is a monolith glass composite consisting of the ionic liquid-entrapped SiO$_2$ network. The ionic liquid, which is hydrophobic in nature, is trapped inside the inorganic silica matrix. The crown ether (DC18C6) is also retained in ionic liquid. The composite glass material was crushed for sorption experiments. Nitrogen adsorption isotherms were determined on a Micromeritics Gemini 2375 surface area analyzer. Mid IR spectra were measured using a Bio-Rad Excalibur FTS 4000 Spectrometer.

Results and Discussion

Figure 1 gives FTIR spectrum of the ionic liquid entrapped inside the silica matrix. The presence of C-H stretching and C-C framework vibrational features in **Fig. 1** indicates the entrapment of the ionic liquid and crown ether.[11] The ionic liquid can be removed from the silica network through solvent extraction using organic solvents.[13] **Figure 2** gives the pore-size distribution of the silica network after removal of the ionic liquid.[13] As seen from **Fig. 2**, the pores formed by the removal of the ionic liquid from the silica network have an average domain size of around 150 Å. Accordingly, the ionic liquid exists inside the sol-gel silica in the form of a nanophasic liquid.

Because DC18C6 is known to selectively complex Sr^{2+},[1, 2] adsorption studies of our composite glass materials were conducted with Sr^{2+} solutions. Strontium-90, which is generated in the nuclear fission process, is similar to calcium and presents a biological hazard because it accumulates in bone tissues. Therefore, various

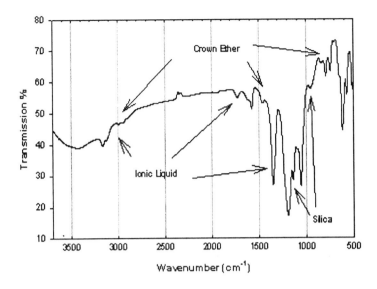

Figure 1 *FTIR spectrum of ionic liquid and crown ether entrapped in silica matrix.*

methods have been developed to extract Sr^{2+} effectively from aqueous solutions. **Table 1** presents Sr^{2+} uptake as a function of the ionic-liquid content. As seen from the table, the sol-gel glass doped with the crown ether but without ionic liquid has a K_d value close to 0.5. The value of K_d sharply increases with the addition of the ionic liquid and levels off at concentrations of approximately 0.6 mL of the ionic liquid in 1 mL of TMOS. This clearly indicates that the ionic liquid plays a key role in the efficient adsorption of Sr^{2+} from the aqueous phase by the composite materials. The greatly enhanced extraction efficiency for the composite materials can be attributed to the unique solvation capability of the room-temperature ionic liquids for the crown ether complex of Sr^{2+}. This also indicates that most of the DC18C6 molecules in the sol-gel glass without the ionic liquid are embedded in the silica matrix and are inaccessible to Sr^{2+}. The incorporation of the nanophase ionic liquid in the silica matrix allows DC18C6 molecules to be solubilized in the ionic liquid instead of the silica network. The diffusion of DC18C6 in the liquid is much faster than that in the solid silica. Therefore, DC18C6 molecules in the ionic liquid are accessible for extraction. In order to determine whether or not the ionicity induced by doping the ionic liquid may contribute to the enhanced extraction, we

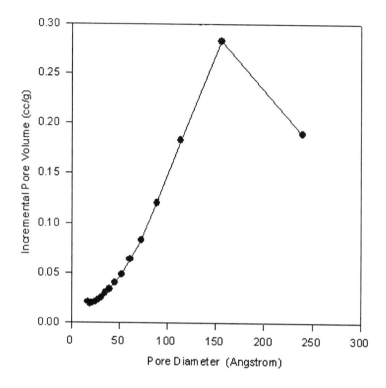

Figure 2 *Pore-size distribution of the silica gel after removal of the ionic liquid. This distribution was calculated from N_2 adsorption isotherm based on BJH (Barret-Joyner-Halenda) method.*

prepared the sol-gel sorbent doped with NaCl instead of the ionic liquid. The rest of the components and sol-gel conditions were kept to be identical. Sorption tests indicated that the affinity ($K_d \approx 1$) of this sorbent toward Sr^{2+} is much less than that of the sorbents doped with the ionic liquid but close to that of the sorbent prepared without the ionic liquid. Accordingly, the ionicity of host matrixes plays no significant role in the enhanced extraction for our composites.

To test the selectivity of our solid-state composite sorbents, we conducted competitive ion-binding experiments using a 1 mM Sr^{2+} solution containing excess amounts of competing ions. The competing ions used in this experiment were Ca^{2+}, Na^+, and Mg^{2+}, which are common in waste sites. The sorbents used in this experiment were prepared using the following composition: 1 mL of ionic liquid, 1 mL of TMOS, 1 mL of formic acid, and 50 mg of DC18C6. The results given in **Table 2** indicate that neither Na^+ or Ca^{2+} interferes with Sr^{2+} extraction, while Mg^{2+} causes some minor interference.

Table 1. The partition coefficient (K_d) as function of the ionic liquid content in sol-gel glasses. The compositions for other sol-gel precursors are 1 ml of TMOS, 1 ml of formic acid, and 50 mg of DC18C6. In all experiments, each sample was equilibrated with 10 mL of $Sr(NO_3)_2$ solution (1×10^{-3} M and pH = 4) in stoppered plastic vials and these mixtures were stirred for an hour at room temperature. The uptake of Sr^{2+} by sorbents was measured with a Perkin Elmer Plasma 400 ICP/AE spectrophotometer.

Ionic liquid used in sol-gel synthesis (ml)	Sr^{2+} uptake %	K_d*
0	10.43	0.5
0.2	47.64	3.6
0.4	92.35	48
0.6	99.10	430
0.8	99.19	480
1.0	99.27	530

*K_d = {(C_i - C_f) /C_f }× {Volume Solution (ml)}/ {Mass Gel (gram)}; where C_i and C_f represent the initial and final solution concentrations, respectively.

Table 2. The partition coefficient (K_d) as function of excess competing metal ions.

Excess competing ions $5x10^{-2}$ M	Sr^{2+} uptake %	K_d
Na^+	97.68	170
Ca^{2+}	98.47	250
Mg^{2+}	82.98	19˙

In conclusion, the hybrid composites consisting of a crown ether, an ionic liquid, and sol-gel silica have been successfully prepared. The ionic liquid, which is hydrophobic and nonvolatile, is trapped in the silica network as a nanophase liquid and provides a diffusion medium for the crown ether molecules. This manifests itself in increased uptake of Sr^{2+} with increased amounts of the ionic liquid. Ionic liquids have been previously used as effective solvents for extraction of organic and inorganic species.[14, 15, 16, -8] The solid-state composite sorbents presented here will make such applications more effective. Currently, we are exploring the use of these sol-gel composites as solid-state membranes for ion-selective electrodes and removal of toxic metal ions.

Acknowledgment: The authors wish to thank Dr. Y. S. Shin and Y. H. Ju for their initial assistance with the synthesis of molten salt precursors. This work was conducted at the Oak Ridge National Laboratory and supported by the Division of Chemical Sciences, Office of Basic Energy Sciences, U.S. Department of Energy, under contract No. DE-AC05-00OR22725 with UT-Battelle, LLC.

References

1. (a) Pedersen, C. J., *J. Am. Chem. Soc.* **1967**, *89*, 7017. (b) Pedersen, C. J., *Angew. Chem. Int. Ed. Engl.* **1988**, *27*, 1021. (c) Cram, D. J. *Angew. Chem. Int. Ed. Engl.* **1988**, *27*, 1009. (d) Lehn, J., *Angew. Chem. Int. Ed. Engl.* **1988**, *27*, 89.

2. (a) Hiraoka, M., "Crown Compounds: Their Characteristics and Applications", Elsevier, Amsterdam, 1982. (b) Lindoy, L. F. "The Chemistry of Macrocyclic Ligand Complexes", Cambridge University Press, Cambridge, 1990. (c) Moyer, B. A. "Complexation and Transport," in *Molecular Recognition: Receptors for Cationic Guests*, Gokel, G. W. Ed., Vol. 1, Comprehensive Supramolecular Chemistry, Atwood, J. L., Davies, J. E. D., MacNicol, D. D.; Vögtle, F.; and Lehn, J. -M. Eds., Pergamon, Elsevier, Oxford, 1996; Chap. 10, pp. 377-416.

3. Izatt, R. M., Powlak, K., Bradshaw, J.S. , Bruening, R. L., *Chem. Rev.* **1991**, *91*, 1721.

4. Kimura, K., Sunagawa, T., Yajima, S., Miyake, S., Yokoyama, M. *Anal. Chem.* **1998**, *70*, 4309.

5. Zhang, B. and Clearfield, A. *J. Am. Chem. Soc.* **1997**, *119*, 2751.

6. (a) Izatt, R. M., Bradshaw, J. S., Bruening, R. L., Bruening, M. L., Tarbet, B. J., Christensen, J. J., *Anal. Chem.* **1988**, *60*, 1826.; (b) Izatt, R. M., Bradshaw, J. S., Bruening, R. L., Bruening, M. L., Tarbet, B. J., Christensen, J. J., *J. Chem Soc, Chem Commun*, **1988**, 812; (c) Izatt, R. M., Bradshaw, J. S., Bruening, R. L., Lamb, J. D., Christensen, J. J., *J. Pure Appl Chem*, **1988**, *60*, 453-460.

7. (a) "Ionic Liquids Prove Increasingly Versatile", *Chemical & Engineering News*, January 4, **1999**, p23-24. (b) "Designer Solvents-Ionic Liquids May Boost Clean Technology Development", *Chemical & Engineering News*, March 30, **1998**, p32-37. (b) Hussey, C. L., *Adv. Molten Salt Chem.*, **1983**, *5*, 185; (c) Hussey, C. L. in *Chemistry of Nonaqueous Solvents*, Popov, A. and Mamantov, G. Editors, Chapter 4, VXH Publishers, New York (1994); (d) Carlin, R. T. and Wilkes, J. S. in *Chemistry of Nonaqueous Solvents*, Popov, A. and Mamantov, G. Editors, Chapter 5, VXH Publishers, New York (1994). (e) Seddon, K. R. *J. Chem. Technol. Biotechnol.* **1997**, *68*, 351. (f) Chauvin, Y. and Olivier-Bourbigou, H. *CHEMTECH* **1995**, *25*, 26.

8. (a) Dai, S., Ju, Y. H., and Barnes, C. E., *J. Chem. Soc. Dalton*, **1999**, 1201. (b) Dai, S., Ju, Y. H., and Luo, H. "Separation of Fission Products Based on Ionic Liquids" in *International George Papatheodorou Symposium*, Patras Science Park, **1999**, p254-262.

9. (a) Carlin, R. T. and Fuller, J. *Chem. Comm.* **1997**, 1345; (b) Fuller, J., Breda, A. C., and Carlin, R. T. *J. Electrochem. Soc.* **1997**, *144*, L67.

10. Kosmulski, M., Osteryoung, R. A., and Ciszkowska, M. *J. Electronchem. Soc.* **2000**, *147*, 1454.

11. Bonhote, P., Dias, A., Papageorigou, N., Kalyansundaram, K., and Gratzel, M. *Inorg Chem*, **1996**, *35*, 1168.

12. Green, W. H., Le, K. P., Grey, J., and Sailor, M. J. *Science*, **1997**, *276*, 1826.

13. Dai, S., Ju, Y. H., Gao, H. J., Lin, J. S., Pennycook, S. J., and Barnes, C. E. *Chem. Comm.* **2000**, 243.

14. Huddleston, J. G.; Willauer, H. D.; Swatloski, R. P.; Visser, A. E.; and Rogers, R. D. *Chem. Comm.* **1998**, 1765.

15. Blanchard, L. A., Hancu, D., Beckman, E. J., Brennecke, J. F. *Nature* **1999**, *399*, 28.

16. Dai, S.; Shin, Y.; Toth, L. M.; and Barnes, C. E. *Inorg. Chem.* **1997**, *36*, 4900.

Chapter 4

Viscosity and Density of 1-Alkyl-3-methylimidazolium Ionic Liquids

Kenneth R. Seddon, Annegret Stark, and María-José Torres

The QUILL Centre, Stranmillis Road, The Queen's University of Belfast, Belfast BT9 5AG, United Kingdom

We report here the viscosity and density of 1-alkyl-3-methylimidazolium salts of $[BF_4]^-$, $[PF_6]^-$, Cl^-, $[CF_3SO_3]^-$ and $[NO_3]^-$. Viscosity decreases as a function of temperature and increases with increasing alkyl chain length, while density decreases with increasing temperature and longer alkyl chain. The viscosity data were fitted to the VFT equation. The ionic liquids were found to exhibit Newtonian behavior when isotropic, whereas they act as non-Newtonian shear–thinning materials at the liquid-crystalline mesophase temperatures.

Introduction

Ionic liquids increasingly gain importance as green solvents (*1,2,3,4,5,6*). In order to implement 1-alkyl-3-methylimidazolium ionic liquids into chemical processes, viscosity and density data are of the utmost importance for chemical engineers.

34

The dramatic effect of traces of water and chloride impurities on the viscosity and density of room-temperature ionic liquids has been reported recently (7). Thus, the results reported here are accompanied by water and chloride measurements. Prior to collecting the viscosity and density data, the ionic liquid samples were dried with heating at *ca.* 70 °C *in vacuo* for 24 h, or until a constant water content was achieved. Water contents were determined by coulometric Karl-Fischer titration (7). Chloride measurements were conducted using a chloride-selective electrode (*ex*: Cole-Parmer) (7).

Viscosity

Viscosity was measured by a LVDV-II Brookfield Cone and Plate Viscometer (1% accuracy, 0.2% repeatability). The sample cup of the viscometer was fitted with luer and purge fittings, so that a positive current of dry dinitrogen was maintained at all times during the measurements, thus avoiding absorption of atmospheric moisture. The sample cup was jacketed with a circulating water bath that was controlled by a circulator bath Grant LTD 6G (± 0.1 °C accuracy).

The viscosity data for $[C_n mim][BF_4]$, $[C_n mim][PF_6]$, $[C_n mim][CF_3SO_3]$, $[C_n mim][NO_3]$ and $[C_n mim]Cl$ (where n is the number of carbons on the 1-alkyl chain of the 1-alkyl-3-*methyl*imidazolium cation), at different temperatures are reported in Tables I, II, III, IV and V.

Table I. Viscosity / cP of $[C_n mim][BF_4]$ ($n = 2$ to $n = 11$) at different temperatures

$T / {}^\circ C$	$n = 2$	$n = 4$	$n = 6$	$n = 8$	$n = 10$	$n = 11$
10.0	110	277	620	897	2071	
20.0	66.5	154	314	439	928	2082
30.0	42.9	91.4	177	240	456	935
40.0	29.4	59.1	106	143	248	473
50.0		39.6	67.9	88.0	147	259
60.0		28.0	45.0	58.0	92.3	152
70.0		20.4	31.5	39.4	60.6	97.4
80.0		15.5	22.6	28.1	42.2	64.9
90.0		11.9	16.6	20.5	27.4	45.5
Water / ppm	525	307	108	80	275	803
% w/w [Cl⁻]	0.03	0.004	0.05	n/a[a]	n/a[a]	n/a[a]

[a] n/a, these ionic liquids are water-immiscible and therefore allow for efficient extraction of chloride salts. Thus, [Cl⁻] is below the detection limit (7).

Table II. Viscosity / cP of $[C_n mim][PF_6]^a$ ($n = 2$ to $n = 10$) at different temperatures

$T/^\circ C$	$n = 2$	$n = 4$	$n = 6$	$n = 8$	$n = 10$	$n = 12$
10.0		728	1434	1848		
20.0		371	690	866		
30.0		204	363	452		
40.0		125	209	252	451	
50.0		82.0	125	152	253	
60.0		55.1	80.3	94.5	150	214
70.0	23.4	37.8	54.0	63.0	96.0	128
80.0	17.1	25.3	37.6	42.8	62.5	80.4
90.0	13.3	19.5	26.8	31.0	43.8	51.0
Water / ppm	n/a[b]	76	28	35	n/a[b]	n/a[b]

[a] these ionic liquids are water-immiscible and therefore allow for efficient extraction of chloride salts. Thus, [Cl⁻] is below the detection limit (7).

[b] Karl-Fischer titration was not performed, since these ionic liquids are solid at room-temperature.

Table III. Viscosity / cP of $[C_n mim][NO_3]$ ($n = 2$ to $n = 8$) at different temperatures

$T/^\circ C$	$n = 2$	$n = 4$	$n = 6$	$n = 8$
10.0		542	1841	2918
20.0		266	804	1238
30.0		144	351	563
40.0	33.8	85.0	190	288
50.0	26.2	54.1	110	159
60.0	18.7	36.9	68.8	95.8
70.0	14.7	25.9	45.4	62.0
80.0	11.3	18.9	31.9	
90.0	8.87	14.6	23.0	30.0
Water / ppm	n/a[a]	1265	441	224
% w/w [Cl]⁻	1.46	0.001	0.047	0.028

[a] Karl-Fischer titration was not performed, since this ionic liquid is solid at room-temperature.

Table IV. Viscosity / cP of [$C_n$mim][CF_3SO_3] ($n = 2$ to $n = 10$) at different temperatures

$T/^{\circ}C$	$n = 2$	$n = 4$	$n = 8$	$n = 10$
10.0	72.2	163	955	2059
20.0	50.0	99.0	492	981
30.0	35.6	64.2	274	512
40.0	26.1	44.1	161	292
50.0	19.5	30.9	101	172
60.0	15.0	22.8	68.3	109
70.0	11.8	17.4	47.3	72.1
80.0	9.45	13.8	34.5	49.9
90.0	7.75	11.0	26.3	35.6
Water / ppm	237	295	98	84
% w/w [Cl]⁻	0.13	0.21	0.014	0.024

Table V. Viscosity / cP of [$C_n$mim]Cl[a] ($n = 2$ to $n = 8$) at different temperatures

$T/^{\circ}C$	$n = 2$	$n = 4$	$n = 6$	$n = 8$
10.0		142000	63060	117000
20.0		40890	18000	33070
30.0		11000	6416	10770
40.0		3800	2543	4524
50.0	341	1534	1124	1930
60.0	173	697	569	927
70.0	99.9	334	311	498
80.0	61.2	182	183	283
90.0	39.1	105	114	172
Water / ppm	5046	1030	7568	3072

[a] Preliminary data only included here; materials with much lower water content are currently being determined.

The viscosity data obtained for [C_2mim][CF_3SO_3] and [C_4mim][CF_3SO_3] are in good agreement with previously published data (8,9). There is also a good agreement with the reported viscosity at 80 °C of [C_2mim][PF_6] (10). The viscosity data for [C_2mim][BF_4] are also in agreement with published data (10,11), although the data are slightly more scattered. This scatter can be due to the hygroscopic nature of [C_2mim][BF_4] and its complete miscibility with water (7), which makes it difficult to obtain chloride and/or water-free. The viscosity data for [C₄mim][BF₄] and [C₄mim][PF₆] reported by Suarez *et al.* (*12*) are clearly higher than those reported here. This difference can be attributed to different measurement methods and/or to the presence of impurities.

Effect of Temperature on Viscosity

The effect of temperature on the viscosity of [C$_n$mim][BF₄] ($n = 2$ to $n = 14$) is displayed in Figure 1. Clearly, for $n < 12$, increasing temperature continuously decreases the viscosity, while [C₁₂mim][BF₄] (□), [C₁₃mim][BF₄] (▲) and [C₁₄mim][BF₄] (■) show a discontinuous viscosity profile. These last-mentioned ionic liquids have been reported to have a liquid crystalline mesophase (*13*), with a lamellar bilayer structure. Thus, the dramatic decrease of the viscosity of [C₁₂mim][BF₄] and [C₁₃mim][BF₄] at 42 °C and 86 °C, respectively, correspond to the transition from the liquid crystal phase to the isotropic disordered liquid (*i.e.* the clearing point). Over the temperature range studied, [C₁₄mim][BF₄] (■) is a liquid crystal and thus shows no clearing point effect on the viscosity. The viscosity of these liquid crystals is described in more detail when discussing the effect of shear rate on the viscosity.

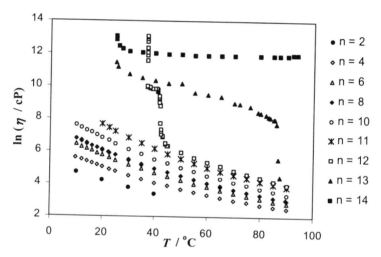

Figure 1. Viscosity of [C$_n$mim] [BF₄] ($n = 2$ to $n = 14$) vs. temperature.

The temperature effect on the viscosity of [C$_n$mim][PF₆] ($n = 2$ to $n = 12$), [C$_n$mim][NO₃] ($n = 2$ to $n = 8$), [C$_n$mim][CF₃SO₃] ($n = 2$ to $n = 10$) and [C$_n$mim]Cl ($n = 2$ to $n = 8$) is shown in Figures 2, 3, 4 and 5 respectively.

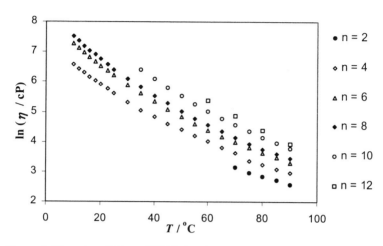

Figure 2. Viscosity of [C$_n$mim] [PF$_6$] (n = 2 to n = 12) vs. temperature.

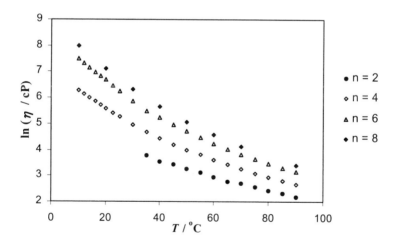

Figure 3. Viscosity of [C$_n$mim] [NO$_3$] (n = 2 to n = 8) vs. temperature.

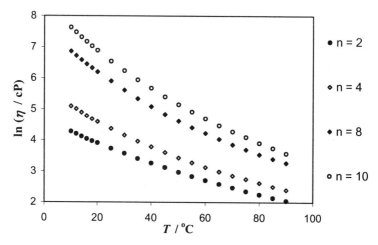

Figure 4. Viscosity of [C$_n$mim] [CF$_3$SO$_3$] (n = 2 to n = 10) vs. temperature.

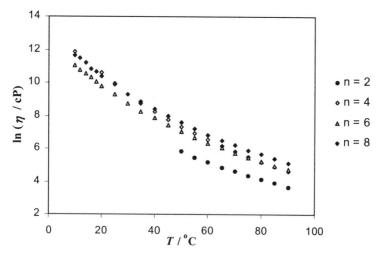

Figure 5. Viscosity of [C$_n$mim] Cl (n = 2 to n = 8) vs. temperature. It should be noted that these materials have a water content > 1000 ppm.

Table VI Least-squares fitted parameters for the VFT equation.

Anion	N	k/K	T_0/K	$ln (A)$	r^2	$\delta^a/10^{-3}$
$[BF_4]^-$	2	996	149.4	-5.5872	0.999994	2
	4	1179	148.4	-5.9607	0.999983	5
	6	1459	142.0	-6.7329	0.999949	9
	8	1468	144.5	-6.6429	0.999978	6
	10	1496	152.9	-6.6922	0.999817	20
$[PF_6]^-$	2	349	243.3	-3.2790	0.999034	10
	4	1712	127.4	-7.2628	0.999386	31
	6	1609	141.4	-6.9126	0.999989	5
	8	1682	139.8	-7.0579	0.999945	11
	10	1804	142.6	-7.3623	0.999941	7
$[NO_3]^-$	2	1447	113.6	-6.5624	0.998329	23
	4	1011	171.0	-5.5444	0.999977	6
	6	1028	182.4	-5.5068	0.999710	26
	8	1325	168.2	-6.3658	0.999860	22
$[CF_3SO_3]^-$	2	1789	73.7	-7.0938	0.999961	5
	4	1054	145.8	-5.4042	0.999933	8
	8	1273	153.4	-5.7660	0.999920	11
	10	1668	139.3	-6.8542	0.999982	6
Cl^-	2	975	208.3	-5.5704	0.999721	14
	4	2232	163.0	-9.4849	0.999667	42
	6	1680	173.4	-7.0795	0.999955	15
	8	1896	167.3	-7.4999	0.999813	32

[a] Fit standard error, $\delta = [\Sigma(ln\ (\eta)_{calcd} - ln\ (\eta)_{expt})^2/(\text{number of points} - \text{number of fitted parameters})]^{1/2}$

The viscosity *vs.* temperature data were least-squares-fitted to the Vogel-Fülcher-Tammann (VFT) empirical equation (*14,15,16*), $\ln(\eta) = k/(T-T_0) + \frac{1}{2}\ln(T) + \ln(A)$ (where T is the absolute temperature), characteristic of glass-forming liquids. The values of the fitted parameters k, T_0 and A are displayed in Table VI.

The empirical parameter T_0 is usually referred to as the "ideal glass transition temperature" and it has been given theoretical significance both through the free volume theory of Cohen and Turnbull (*17*) and the configurational entropy approach of Adam and Gibbs (*18*). Below T_0 the material exists in a state of closest random packing of the molecules or ions as an equilibrium glass with zero mobility (*19*). Experimentally, this thermodynamic equilibrium cannot be attained, and the experimental glass transition, T_g, is observed instead, as a sudden change in properties such as heat capacity, volume or viscosity. Thus, for the VFT parameter T_0 to represent a theoretically meaningful ideal glass transition temperature, it has to be lower than the observed T_g. The values of T_0 obtained in this study for [$C_n$mim][BF_4] and [$C_n$mim][PF_6] are all lower than the reported T_g values for such systems (*13,20*) by 50-60 °C.

For smaller temperature ranges, the data may also be fitted to the Arrhenius equation. However, the analysis of the residual errors, r^2 and δ showed that the VFT equation describes the temperature behavior more accurately, and is thus more versatile. The kinematic and absolute viscosity of 1,3-dialkylimidazolium chloroaluminate ionic liquids (*21*) (temperature range –10 to 95 °C) and for [C_2mim]Br-AlBr$_3$ (*22*) (temperature range *ca.* 25-100 °C) has been reported previously to follow non-Arrhenius behavior. In both studies, the viscosity was found to fit the VFT equation better than the Arrhenius equation. The non-linearity of the Arrhenius plots of the viscosity of [C_2mim][CF_3SO_3] (*8,9*), [C_4mim][CF_3SO_3], [C_2mim][Tf_2N] (where [Tf_2N]$^-$ is [bis(trifluoro-methylsulfonyl)amide]$^-$) and [C_4mim][Tf_2N] (*8*) (temperature range 5 to 85 °C) has also been reported. On the other hand, Arrhenius behavior has been reported for the absolute viscosity of *N*-alkylpyridinium chloroaluminate (*23*) ionic liquids (temperature range 25 to 75 °C) and for the kinematic viscosity of [C_2mim][BF_4] (*11*) (temperature range 20 to 100 °C).

Effect of the alkyl chain length

The viscosity of the 1-alkyl-3-methylimidazolium ionic liquids was found to increase with increasing alkyl chain length, although this increase is not linear (see Figure 6). The increasing viscosity of ionic liquids with longer alkyl chain lengths has been reported previously (*8,21,23*).

Longer alkyl chain lengths in the 1-alkyl-3-methylimidazolium cation not only lead to heavier and bulkier ions, but also give rise to increasing Van der Waals attractions between the aliphatic alkyl chains (*8*). Van der Waals forces occur increasingly, for $n \geq 4$, since the ionic headgroup has little or no electron-withdrawing effect on that part of the alkyl chain. In the extreme case, with n >12, liquid crystal phases are formed with a microphase separation of the hydrophilic ionic headgroups and the hydrophobic alkyl chains. This scenario is consistent with the interdigitated layered structure observed in the solid state for long chain ionic liquid like [C_{12}mim][PF_6] (*20*).

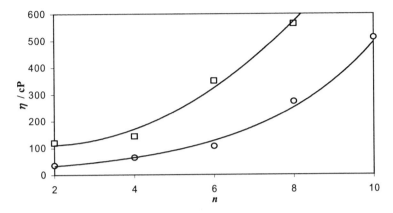

Figure 6. Viscosity at 30 °C of [$C_n$mim] [NO_3] (□) and [$C_n$mim] [CF_3SO_3] (O) vs alkyl chain length.

Effect of the shear rate

The effect of shear rate (the velocity gradient within the flowing liquid) on viscosity indicates whether a liquid is Newtonian or non-Newtonian. Newtonian liquids display a viscosity which is independent of the shear rate (24). The viscosity of [$C_n$mim][BF_4] (n = 4 to n = 8) and [$C_n$mim][PF_6] (n = 4 to n = 12) ionic liquids were tested for Newtonian behavior by measuring the viscosity at different shear rates. The same experiment was performed on [C_{12}mim][BF_4] at 40 °C, in the liquid crystal phase, and at 65 °C, when the ionic liquid is an isotropic liquid. For [$C_n$mim][BF_4] (n = 4 to n = 8) and [$C_n$mim][PF_6] (n = 4 to n = 12) and for [C_{12}mim][BF_4] at 65 °C, the viscosity values at the different shear rates were constant within experimental error for all these ionic liquids. The results for [C_{12}mim][BF_4] at 65 °C are displayed in Figure 7 (top). In contrast, Figure 7 (bottom) shows how the viscosity of the liquid crystalline [C_{12}mim][BF_4] decreases as the shear rate increases. This is typical non-Newtonian shear-thinning behavior and reflects the breaking up of the layered structure of this liquid crystal by the shearing action. Furthermore, when the shear rate was reduced, the viscosity values were lower than the ones obtained on the increasing shear rate cycle. This indicates that the structure of the liquid crystal cannot be totally recovered, at least on the time scales used (1 h for the whole experiment, 330 sec between speed increases / decreases).

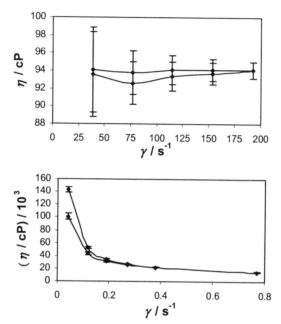

Figure 7. Viscosity of [C₁₂mim] [BF₄] at 65 °C (top) and 40 °C (bottom) vs. shear rate. The upper lines in both graphs correspond to increasing shear rate (i.e. from low shear rates to high shear rates) and the lower lines correspond to decreasing shear rate (i.e. from high shear rates to low shear rates)

Density

Density was measured using calibrated 10 cm³ density bottles, with an experimental error of ± 0.0008 g cm⁻³. The density bottles were immersed in a water bath (Grant LTD 6G, ±0.1°C accuracy). The density data for [$C_n$mim][BF₄], [$C_n$mim][PF₆], [$C_n$mim][NO₃] and [$C_n$mim]Cl as a function of temperatures are reported in Tables VII, VIII, IX, and X. Chloride and water contents are the same as reported in the viscosity tables (Tables I, II, III and V). The density data obtained for [C_2mim][BF₄] agree well with previously reported data (*11*), while the data for [C_4mim][BF₄] and [C_4mim][PF₆] are again slightly lower than that reported by Suarez *et al.* (*12*).

Table VII. Density / g cm⁻³ of [C$_n$mim][BF$_4$] (n = 2 to n = 10) at different temperatures

$T/°C$	$n = 2$	$n = 4$	$n = 6$	$n = 8$	$n = 10$
20.0	1.2479	1.2077	1.1531	1.1095	1.0723
30.0	1.2401	1.2017	1.1455	1.1001	1.0670
40.0	1.2334	1.1947	1.1382	1.0933	1.0603
50.0	1.2262	1.1885	1.1323	1.0868	1.0533
60.0	1.2181	1.1807	1.1248	1.0801	1.0461
70.0	1.2102	1.1748	1.1156	1.0730	1.0395
80.0	1.2050	1.1676	1.1095	1.0657	1.0330
90.0	1.1972	1.1603	1.1031	1.0574	1.0272

Table VIII. Density / g cm⁻³ of [C$_n$mim][PF$_6$] (n = 4 to n = 8) at different temperatures

$T/°C$	$n = 4$	$n = 6$	$n = 8$
20.0	1.3727	1.3044	1.2345
30.0	1.3626	1.2920	1.2220
40.0	1.3565	1.2798[a]	1.2124
50.0	1.3473	1.2681	1.2013[d]
60.0	1.3359	1.2570	1.1929[e]
70.0	1.3285	1.2521[b]	1.1877[f]
80.0	1.3225	1.2389[c]	1.1753
90.0	1.3126	1.2296	1.1676

[a] 37.0 °C; [b] 67.0 °C; [c] 82.0 °C; [d] 55.0 °C; [e] 64.0 °C; [f] 72.0 °C

Table IX. Density / g cm⁻³ of [C$_n$mim][NO$_3$] (n = 4 and n = 6) at different temperatures

$T/°C$	$n = 4$	$n = 6$	$T/°C$	$n = 4$	$n = 6$
20.0	1.1574	1.1185	60	1.1309	1.0951
30.0	1.1497	1.1144	70	1.1239	1.0892
40.0	1.1435	1.1078	80	1.1167	1.0820
50.0	1.1372	1.1014	90	1.1120	1.0769

Table X. Density / g cm⁻³ of [C$_n$mim]Cl^a (n = 6 and n = 8) at different temperatures

T / °C	$n = 6$	$n = 8$
25.0	1.0338	1.0124
40.0	1.0241	1.0074
60.0	1.0135	0.9999

^{a} Preliminary data only included here; materials with much lower water content are currently being determined

The decrease in the density of ionic liquids with increasing temperature can be fitted to linear equations of the form $\rho = a + b$ (T-60), where T is the temperature in °C, a is the density at the arbitrary temperature, 60 °C, in g cm⁻³, and b is the temperature coefficient in g cm⁻³ K⁻¹. The fitting values obtained for a and b for density of all the studied ionic liquids are tabulated in Table XI. The trends are displayed graphically in Figure 8 for [C$_n$mim][BF$_4$] (n = 2 to n = 10).

The effect of the alkyl chain length on the density of ionic liquids is also linear (see Figure 9). The decrease of density with increasing size was reported previously for the 1,3-dialkylimidazolium chloroaluminate ionic liquids by Wilkes and co-workers (21), and the N-alkylpyridinium chloroaluminate ionic liquids by Hussey and co-workers (23). The same feature is reflected in the density data reported by Bonhôte et al. (8).

Table XI. Least-squares fitted parameters a / g cm⁻³ and b / g cm⁻³ K⁻¹

Anion	n	a	$-b / 10^{-4}$	r^2	$\delta^a / 10^{-4}$
[BF$_4$]⁻	2	1.2187	7.2359	0.9987	7
	4	1.1811	7.6229	0.9992	5
	6	1.1242	7.2090	0.9982	8
	8	1.0796	7.1304	0.9979	8
	10	1.0466	6.6034	0.9989	6
[PF$_6$]⁻	4	1.3381	8.5275	0.9962	14
	6	1.2596	10.2938	0.9921	25
	8	1.1960	9.2302	0.9928	21
[NO$_3$]⁻	4	1.1306	6.5239	0.9985	6
	6	1.0951	6.1377	0.9980	7
Cl⁻	6	1.0132	5.7756	0.9973	7
	8	0.9999	3.6033	0.9990	3

^{a} Fit standard error, $\delta = [\Sigma(\rho_{calcd} - \rho_{expt})^2 / (\text{number of points-number of fitted parameters})]^{1/2}$

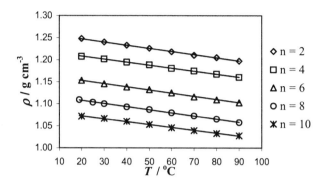

Figure 8. Density of [C$_n$mim] [BF$_4$] (n = 2 to n = 10) vs. temperature.

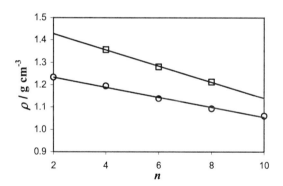

Figure 9. Density of [C$_n$mim] [BF$_4$] (n = 2 to n = 10) (O) and [C$_n$mim] [PF$_6$] (n = 2 to n = 8) (□) at 40 °C vs. alkyl chain length.

48

Conclusions

We have found suitable equations to predict the viscosity and density of an array of ionic liquids as a function of temperature and alkyl chain length. With regards to temperature, the viscosity data fit the VFT equation for glass forming liquids. Lengthening the alkyl chain results in an increased viscosity. Liquid-crystalline ionic liquids show a substantially higher viscosity than isotropic ionic liquids and behave as non-Newtonian shear-thinning materials. Density decreases linearly with both alkyl chain length and temperature. Because viscosity and density behave in a regular and predictable manner as a function of chain length and temperature, these properties can be easily modeled for chemical engineering purposes.

Acknowledgements

We (AS and M-J T) would like to thank the European Union for the funding of this study which was conducted within the BRITE-EURAM III framework (Contract no BRPR-CT97-043, Project no BE96-3745), and the EPSRC and Royal Academy of Engineering (KRS) for funding a clean technology fellowship. We are also indebted to Dr. B. Ellis (BP Amoco), Dr. J.D. Holbrey, and to the Dept. of Chemical Engineering of the Queen's University of Belfast. We also wish to thank Dr. R. Wareing (Elementis) for the supply of imidazolium chlorides.

References

1. Rooney, D. W.; Seddon, K. R. In *Handbook of solvents;* Wypych, G., Ed.; William Andrew: Toronto, 2000, pp 1459-1484.
2. Earle, M. J.; Seddon, K. R. *Pure Appl. Chem.* **2000,** *72*, xxxx-xxxx.
3. Earle, M. J.; Seddon, K. R. In *Clean Solvents;* Abraham, M. and Moens, L., Ed.; ACS Symp. Ser., 2000, pp in press.
4. Welton, T. *Chem. Rev.* **1999,** *99*, 2071-2083.
5. Freemantle, M. *Chem. & Eng. News* **2000,** *78*, 15th May, 37-50.
6. Holbrey, J. D.; Seddon, K. R. *Clean Prod. Proc.* **1999,** *1*, 223-236.
7. Seddon, K. R.; Stark, A.; Torres, M. J. *Pure Appl. Chem.* **2000,** *72*, 12, 2275-2287.
8. Bonhôte, P.; Dias, A. P.; Armand, M.; Papageorgiou, N.; Kalyanasundaram, K.; Grätzel, M. *Inorg. Chem.* **1996,** *35*, 1168-1178.

9. Cooper, E. I.; O'Sullivan, E. M. In *New, stable, ambient-temperature molten salts;* ; Electrochem. Soc.: Pennington, NJ, 1992; Vol. 92, pp 386-396.

10. McEwen, A. B.; Ngo, H. L.; LeCompte, K.; Goldman, J. L. *J. Electrochem. Soc.* **1999,** *146,* 1687-1695.

11. Fuller, J.; Carlin, R.; Osteryoung, R. A. *J. Electrochem. Soc.* **1997,** *144,* 3881-3886.

12. Suarez, P. A. Z.; Einlof, S.; Dullius, J. E. L.; de Souza, R. F.; Dupont, J. *J. Chim. Phys.* **1998,** *95,* 1626-1639.

13. Holbrey, J. D., Seddon, K.R. *J. Chem. Soc., Dalton Trans.* **1999,** 2133-2139.

14. Tammann, G.; Hesse, W. *Z. anorg. allg. Chem.* **1926,** *156,* 245-257.

15. Fulcher, G. S. *J. Am. Ceram. Soc.* **1925,** *8,* 339-355.

16. Vogel, H. *Phys. Z.* **1921,** *22,* 645-646.

17. Cohen, M. H.; Turnbull, D. *J. Chem. Phys.* **1961,** *34,* 120-125.

18. Adam, G.; Gibbs, J. H. *J. Chem. Phys.* **1965,** *43,* 139-146.

19. Angell, C. A.; Moynihan, C. T. In *Molten Salts;* Mamantov, G., Ed.; Marcel Dekker: New York, 1969, pp 315-375.

20. Gordon, C. M.; Holbrey, J. D.; Kennedy, A. R.; Seddon, K. R. *J. Mater. Chem.* **1998,** *8,* 2627-2636.

21. Fannin, A. A.; Floreani, D. A.; King, L. A.; Landers, J. S.; Piersma, B. J.; Stech, D. J.; Vaughn, R. L.; Wilkes, J. S.; Williams, J. L. *J. Phys. Chem.* **1984,** *88,* 2614-2621.

22. Sanders, J. R.; Ward, E. H.; Hussey, C. L. *J. Electrochem. Soc.* **1986,** *133,* 325-330.

23. Carpio, R. A.; King, L. A.; Lindstrom, R. E.; Nardi, J. C.; Hussey, C. L. *J. Electrochem. Soc.* **1979,** *126,* 1644-1650.

24. Barnes, H. A.; Howard, A. *An introduction to rheology;* Elsevier: Amsterdam, 1989.

Chapter 5

Reusable Reaction Systems Derived from Fluorous Solvents or Ionic Liquids and Catalysts

Tomoya Kitazume

Graduate School of Bioscience and Biotechnology, Tokyo Institute of Technology, Nagatsuta, Midori-ku Yokohama 226–8501, Japan

Construction and behaviors of reusable reaction systems derived from fluorous solvents or ionic liquids and catalysts, including the discovery and development of fluorous solvent-metal catalyst and ionic liquid-Ln(OTf)$_3$, are described based on the idea of green chemistry.

Studies of reaction media and conditions for improved selectivity and energy minimization to decrese the damage for the endocrine system are having an important impact on organic reactions. In general, fluorous (perfluorinated) fluids such as perfluoroalkanes, perfluoroalkyl ethers and perfluoroalkylamines, have been recognized to have extremely stable, nonpolar/inert characteristics in organic reactions, and they are suitable for reaction containing unstable reagents, for heat transfer, and for refluxing and for separating a low boiling components.[1] Their fluids all have very unusual properties, such as high density and high stability, low solvent strength, and extremely low solublity in water and organic materials.[2]

As mentioned above, reuse of solvents is favorable for economical and environmental reasons. Ionic liquids, with low viscosity and no measurable vapour pressure, can be used as environmentally benign organic media for a broad spectrum of chemical processes.

Therefore, the first issue is the scope and limitations of the utility of fluorous fluids as alternative reaction media for a variety of reactions, which are of great current interest due to their unique reactivities and selectivities. The construction of new reusable reaction system made from ionic liquid-organomolecular catalyst is the next.

The Utility of Fluorous Fluids as Alternative Reaction Media

As fluorous (perfluorinated) fluids have been recognized as new alternative solvents,[4,5] initially, we examined Hosomi-Sakurai reaction[6] as a model reaction and results are summarized in Table 1. In this reaction, we have found that perfluorotrialkylamines were better reaction media than perfluorohexane. In addition, perfluorotriethylamine was superior to perfluorotri-n-butylamine presumably due to higher lipid solubility derived from shorter perfluoroalkyl chains, and turned out to be as an appropriate solvent as CH_2Cl_2 which is commonly used in this reaction (entry 3 vs 4).

$$RCHO \; + \; \diagup\!\!\!\diagdown\!\!\!\diagup TMS \; \xrightarrow{\;a\;} \; R\diagup\!\!\!\!\!\overset{OH}{\diagdown}\!\!\!\diagup\!\!\!\diagdown$$

^{a}TiCl$_4$ (1.0 eq), 15 min, -78°C $\longrightarrow$ rt

Table 1. The Hosomi-Sakurai allylation

Entry	R	Solvent	Yield (%)	Entry	R	Solvent	Yield (%)
1	n-C$_5$H$_{11}$	n-C$_6$F$_{14}$	67	5	n-C$_5$H$_{11}$	Hexane	63
2		(C$_4$F$_9$)$_3$N	78	6		Neat	67
3		(C$_2$F$_5$)$_3$N	90	7	PhCH$_2$CH$_2$	(C$_2$F$_5$)$_3$N	98
4		CH$_2$Cl$_2$	92	8	c-C$_6$H$_{11}$	(C$_2$F$_5$)$_3$N	quant.

Yields were obtained when the reaction was carried out in nonpolar hexanes (entry 5) or under neat condition (entry 6). These results shows that some solvent effects of perfluorotriethylamine promoted the above reaction. Some comments are worth noting; e.g., (1) the basicity of perfluorotrialkylamine is so reduced that the amine does not interact with such a strong Lewis acid like TiCl$_4$, (2) however, the amine's polarity might be retained to some extent[7] and it might be an important factor of the proceeding of reactions as well as lipophilicity. Althogh lipid solubility of perfluorotri-*n*-butylamine is lower than perfluorohexane and hexanes, it was found to be a better reaction medium. Among fluorous amines, perfluorotriethylamine which shows relatively high lipophilicity due to short perfloroalkyl chains and appropriate boiling point, could be an ideal altarnative solvent.

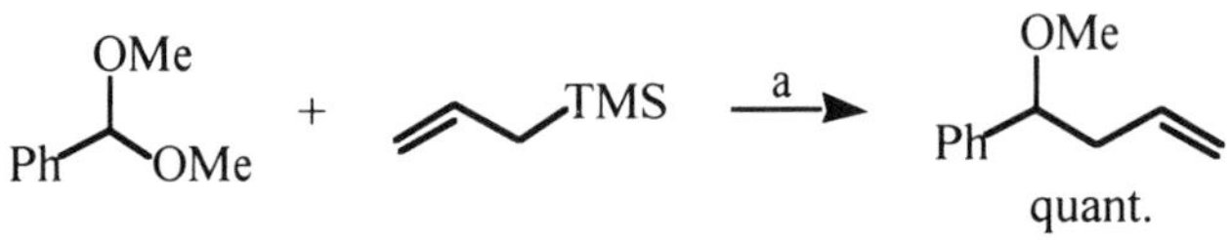

$$^a \text{TMSOTf (10 mol \%), } (C_2F_5)_3N, \ 0 \ °C \longrightarrow \text{ rt. 1h.}$$

Figure 1.   Allylation in perfluorotriethylamine.

A cataliytic system of this reaction[8] using acetal in place of aldehyde proceeded smoothly (Figure 1). While perfluorotriethylamine shows some miscibility with organic materials, extractive workup with ethylacetate/water resulted in high recovery of the reaction medium (90%). Successive reuse of the recovered solvent in the same reaction without further purifcation yielded amounts of product as high as in the first cycle (Figure 2: Yield. Cycle 1 : 90%; cycle 2 : 92%). This result clearly shows that high hydrophobic character of perfluorinated material allows us to recycle directly the reaction medium for highly moisture-sensitive reaction.

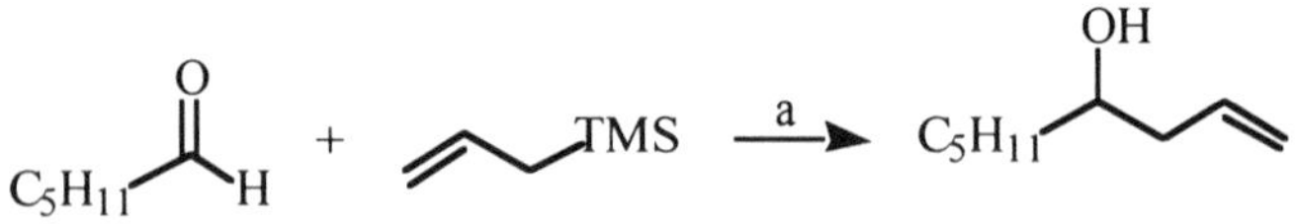

$$^a \text{TiCl}_4 \ (1.0 \ \text{eq}), \ (C_2F_5)_3N$$

Figure 2. Recycling perfluorotriethylamine

Friedel-Crafts Reaction in Fluorous Fluids

It is well known that Friedel-Crafts reaction is one of the most important synthetic reactions especially in industrial production,[9] however, this reaction is usually carried out in toxic and/or harmful organic reaction media like CH_2Cl_2, CS_2 etc. Recently, it has been reported that benzotrifluoride (BTF) was a useful alternative solvent of CH_2Cl_2 and Friedel-Crafts acylation was performed.[10] However, BTF reacts with $AlCl_3$ which is typically used in Friedel-Crafts reactions (Scheme 1). In addition, BTF is sensitive to some kind of reducing conditions and hydrolyzed by aqueous acid at high temperature.[7] These reactivities clearly limit its utility as a reaction medium.

Scheme 1

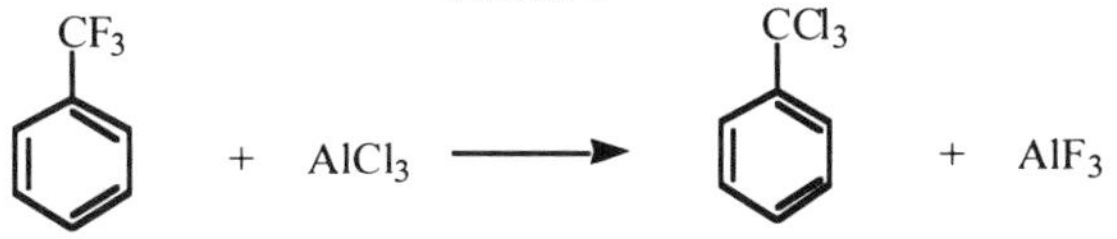

In the next step, we would like to describe the utility of fluorous fluids as the reaction media for Friedel-Crafts reaction. We have studied the aluminum chloride-catalyzed Friedel-Crafts reaction between aromatic compounds and acetyl chloride. As the results shown in Table 2, the catalyzed acetylation of benzene and p-xylene using an equimolar amount of $AlCl_3$, was performed at

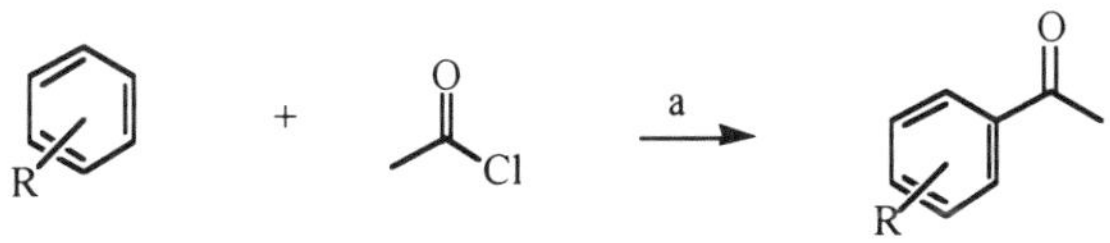

^{a} $AlCl_3$ (1.0 eq.), rt, overnight

Table 2.   Results of the Friedel-Crafts reaction with $AlCl_3$

Entry	Arene	Solvent	Yield (%)	Entry	Arene	Solvent	Yield (%)
9	Benzene	$(C_2F_5)_3N$	89	14	p-Xylene	$2\text{-}n\text{-}C_4F_9\text{-}F\text{-}THF$	46
10		$(C_4F_9)_3N$	81	15		$n\text{-}C_6F_{14}$	63
11		$2\text{-}n\text{-}C_4F_9\text{-}F\text{-}THF$ $^{a)}$	87	16		CH_2Cl_2	78
12	p-Xylene	$(C_2F_5)_3N$	55	17	Anisol	$(C_2F_5)_3N$	89 $^{b)}$
13		$(C_4F_9)_3N$	57	18	Mesitylene	$(C_2F_5)_3N$	quant.

^{a} perfluoro-2-butyltetrahydrofuran ^{b} only p-adduct was obtained

room temperature by employing perfluorotriethylamine as the solvent, resulting in 89 and/or 55% yields, respectively; however the reaction in hexane did not proceed. From the result of the reusable fluorous solvents, fluorous liquids possess the possibility to be good reaction media, especially for Friedel-Crafts reaction. Common problems in the use of stoichiometric amount of AlCl$_3$, are its instability and disposal of stoichiometric amount of Al(OH)$_3$ after aqueous work-up. In view of 'Atom Economy',[11] catalytic reactions are preferable.

OMe + PhCOCl $\xrightarrow{a}$ OMe COPh

t-Bu   t-Bu

^{a} ZnCl$_2$ (10 mol %), reflux, 40 h

Table 3. Catalytic acylation with ZnCl$_2$

Entry	Solvent	Bath temp. (° C)	Yield (%)
19	sym-Tetrachloroethane	138	66
20	Perfluoro-2-butyltetrahydrofuran	99-107	62
21	Perfluorotriethylamine	70	64

Hence, catalytic acylation with ZnCl$_2$, was next investigated (Table 3). In the literature, this benzoylation was performed at reflux temperature in highly toxic sym-tetrachloroethane (entry 19). The same reaction was carried out in perfluoro-2-butylterahydrofran under the same condition (entry 20) and benzoylated product was isolated in comparable yield. In entry 21, the same reaction was successfully carried out in perfluorotriethylamine at a lower reflux temperature. These results show that fluorous media with lower reflux temperatures and non-flammability are good substitutes in Friedel-Crafts acylation for conventional organic liquids.

The use of catalysts bearing perfluorinated ligands and their recovery after reaction using a fluorous biphase system has been of great concern to chemists. However, such 'fluorous catalysts' are not commercially available and their synthesis often requires tedious steps and expensive starting materials. Therefore, before pursuing perfluorinated catalysts, we tried to utilize commercially available catalysts which can be recycled easily. Sc(OTf)$_3$ has been shown to be a good Friedel-Crafts catalyst and can be recovered quantitatively after extractive work-up in aqueous phase separated from organic products. As shown in Figure 3, Sc(OTf)$_3$ effectively catalyzes acetylation of anisole in perfluorinated solvents and only p-adduct was obtained in 69 %, 3 times the average isolated yield.

Moreover, benzaldehyde dimethylacetal was also reacted with anisole to produce a disubstituted material in 57% yield. Further, $Sc(OTf)_3$ can be recovered free from organic products by simple extractive work-up. Successive reuse of the recovered $Sc(OTf)_3$ and solvent in the same reaction without further purification yielded the product in 40% yield.

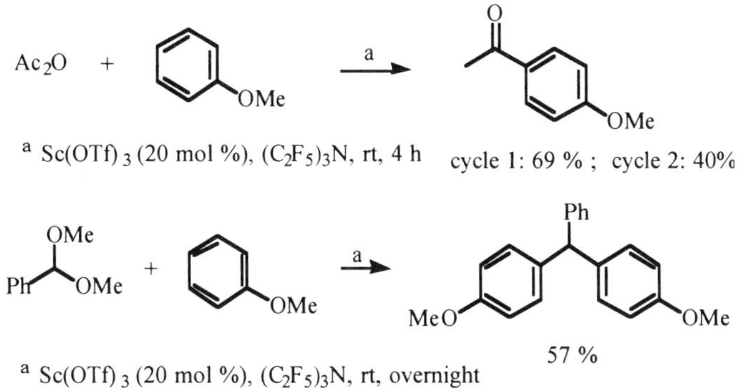
Ac2O
+
OMe
a
O
OMe
a Sc(OTf) 3 (20 mol %), (C2F5)3N, rt, 4 h
cycle 1: 69 % ; cycle 2: 40%
OMe
Ph
OMe
+
OMe
a
Ph
MeO
OMe
57 %
a Sc(OTf) 3 (20 mol %), (C2F5)3N, rt, overnight

Figure 3. Reusable Sc(OTf) 3 - (C2F5)3N System

The Construction of Reusable Ionic Liquid-Catalyst System

The ionic liquids are low-melting point molten salts that have a liquid range of about 300 °C. Further, these ionic organic liquids, which in some cases can serve as both catalyst and solvent, are attracting attention from chemical processes. In our continuous study of the reusable reaction media, we have examined the construction of new type of reusable reaction system made from ionic liquid and catalyst.

Ionic liquid-$M(OTf)_3$ reaction syste

Aza-Diels-Alder reaction is a well known synthetic reaction in the synthesis of azasugars and their derivatives, which often exhibit unique physiochemical properties.[12] Initially, we have prepared the new ionic liquids (such as 8-ethyl-1,8-diaza-bicyclo-[5,4,0]-7-undecenium trifluoromethanesulfonate [EtDBU][OTf] and 8-methyl-1,8-diazabicyclo-[5,4,0]-undecenium trifluoromethanesulfonate [MeDBU][OTf]) directly from the reaction of 1,8-diaza-bicyclo[5,4,0]-7-undecene with ethyl or methyl trifluoromethanesulfonate, yielding in 98% and 95%, respectively. (Figure 4)

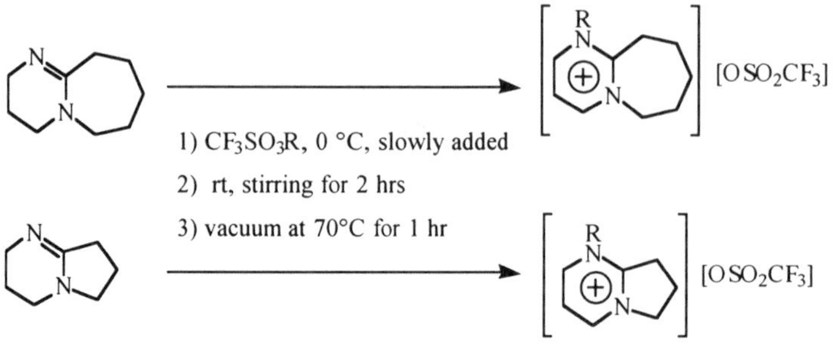

Figure 4. Preparation of ionic liquids [EtDBU][OTf] and [MeDBU][OTf]

We carried out the *aza*-Diels-Alder reaction of *N*-phenyl phenyl imine with 1-methoxy-3-(trimethylsilyl)oxy-1,3-butadiene in an ionic liquid [EtDBU][OTf]. In this reaction, the uncatalysed reaction did not proceed. The addition of Lewis acid (microencapsulated scandium trifluoromethanesulfonate, Wako Pure Chemical Industries, Ltd.) dramatically increases the yield to 75% and 67%, respectively. The present reaction summarized in Table 4 smoothly proceeded at room temperature in an ionic liquid.[13]

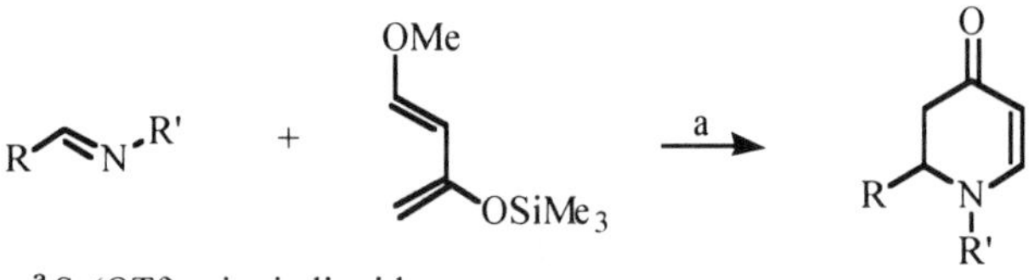

^{a} Sc(OTf)$_3$, ionic liquid

Table 4. Synthesis of 5,6-dihydro-4-pyridones

entry	R	R'	Ionic liquid $^{a)}$	Yield (%)	entry	R	R'	Ionic liquid $^{a)}$	Yield (%)
1	Ph	Ph	1	75	3	Ph	3,4-F$_2$C$_6$H$_3$	1	74
2	Ph	Ph	2	67	4	4-CF$_3$C$_6$H$_4$	Ph	1	25

^{a} **1** [EtDBU][OTf] ; **2** [emim][OTf]

In the next satep, we have examined a second version of this tandem Mannich-Michael-type reaction which can be carried out on a one-pot reaction which is preferable in view of 'green chemistry'. In this reaction system shown in Table 5, initially the corresponding imines were prepared *in situ* from the reaction of aldehyde and amine in an ionic liquid, and then Lewis acid and 1-methoxy-3-(trimethylsilyl)oxy-1,3-butadiene were added into the above reaction mixture. After 20 h of stirring at room temperature, N-aryl-6-aryl-5,6-dihydro-4-pyridone was obtained by the extraction with crude diethyl ether, and ionic liquid ([EtDBU][OTf]) and microencapsulated Lewis acid were recovered more than 90% yield. Before the use and reuse of ionic liquids, ionic liquids were purified under dynamic vacuum at 70-80 °C for 1 h, and then we checked the purity and structure by the ^1H and ^{19}F NMR specta (no other peaks except an ionic liquid). Successive reuse of the recovered ionic liquids and microencapsulated Lewis acid in the same reaction yielded amounts of product as high as in the first cycle shown in Table 5. In the third cycle, reuse of ionic liquids and microencapsulated Lewis acid recovered from the second cycle is possible to produce the same 5,6-dehydroxy-4-pyridone in the same reaction. Quenching after the third cycle, more than 90% of an ionic liquid [EtDBU][OTf] was recovered (Table 5)..

^{a} ionic liquid^b CH$_2$=C(OSiMe$_3$)CH=CHOMe, Sc(OTf)$_3$, ionic liquid

Table 5. One-pot synthesis of 5,6-dihydro-4-pyridone

Entry	R	R'	Ionic Liquid$^{a)}$	Yield $^{b)}$ (%)	Entry	R	R'	Ionic Liquid$^{a)}$	Yield $^{b)}$ (%)
1	Ph	Ph	1	82$^{c)}$	8	Ph	4-FC$_6$H$_4$	1	75
2			2	80 $^{c)}$	9			1	82 $^{d)}$
3	3,4-F$_2$C$_6$H$_3$	1		95	10			1	95 $^{e)}$
4			2	99	11	4-FC$_6$H$_4$	Ph	1	85
5			2	99$^{d)}$	12			2	95
6			2	99 $^{e)}$	13	4-CF$_3$C$_6$H$_4$		1	85
7	4-FC$_6$H$_4$		2	79	14			2	88

^{a} ionic liquid:**1** [EtDBU][OTf] ; **2** [emim][OTf] ^{b} Yields were determined by ^{19}F NMR integral intensities ^{c} isolated yiled ^{d} second cycle ^{e} third cycle

Ionic liquid ([EtDBU][OTf]) is also a reusable media in the preparation of heterocycles shown inTable 6.The reaction of benzaldehyde with 2-aminobenzyl alcohol in ionic liquid ([EtDBU][OTf]), smoothly proceeded at room temperature. The produced 4H,2H-2-phenyl-3,1-benzoxazine was separated by the extraction with diethyl ether, and ionic liquid ([EtDBU][OTf]) was recovered more than 98%. Furthermore, reuse of the recovered Ionic liquid ([EtDBU][OTf]) in the same reaction yielded amounts of product as high as in the first cycle shown in Table 6.[14] When we used tetrahydrofuran as a solvent in the above reaction (91% yield) to compare the solvent effect, it is not easy to separate the solvent and product by the direct distillation. If the water was used for quenching the reaction, tetrahydrofuran is not reuse in the same reaction.

Table 6.   Synthesis of heterocycles in an ionic liquid

Entry	Arene	RCHO R	Ionic Liquid[a]	Yield (%)	Entry	Arene	RCHO R	Ionic Liquid[a]	Yield (%)
1	CH$_2$OH / NH$_2$	C$_6$H$_5$	1	95	11	Cl / CH$_2$OH / NH$_2$		1	96
2			1	97[b]	12		C$_6$H$_5$	2	84
3			1	97[c]	13			3	>99
4			2	89	14			4	98
5			3	98	15			5	97
6			4	98	16	CH$_2$NH$_2$ / NH$_2$		1	92
7			5	98	17			2	94
8		4-FC$_6$H$_4$	3	97	18			3	>99
9			3	96[b]	19			4	96
10		4-CF$_3$C$_6$H$_4$	3	95	20			5	94

[a] ionic liquid:**1** [EtDBU][OTf] ; **2** [MeDBU][OTf] ; **3** [emim][OTf] ;
4 [bmim][BF$_4$] ; **5** [bmim][PF$_6$]   [b] second cycle [c] third cycle

A Lewis Acid-Catalysed Sequential Reaction in Ionic Liquids

In the next stage, we would like to describe the possiblity of reusable of ionic liquid at 200 °C, proceeding one of sequential syntheses (domino reactions) containing the Claisen rearrangement[15] and cyclization reactions in the presence of Lewis acid to give 2-methyl-2,3-dihydro-benzo[b]furan derivatives. The Claisen rearrangement of allyl phenyl ether is carried out at 150 °C for 4 h in an ionic liquid ([EtDBU][OTf]). In this system, the uncatalysed reaction did not proceed. The addition of Lewis acid (Sc(OTf)$_3$) promotes the Claisen rearrangement and cyclization reactions after 4 h at 200 °C, 2-allyl phenol and 2-methyl-2,3-dihydro-benzo[b]furan and the starting material were detected by

^{1}H NMR spectra in the ratio of 1:1:1. When the reaction was carried out at 200 °C for 10 h, the expected rearrengemnt material, 2-allylphenol was not detected. The obtained product is 2-methyl-2,3-dihydrobenzo[b]furan which is produced from the domino reaction containing the Claisen rearrangement and intramolecular cyclization of 2-allyl phenol catalysed by Lewis acid (Table 7). To study the reaction mechanism, we examined the cyclization reaction of 2-allyl phenol in the system of Sc(OTf)$_3$ and ionic liquid ([EtDBU][OTf]) under the same reaction condition. After 4 h of heating, the target material, 2-methyl-2,3-dihydrobenzo[b]furan was obtained in 61% yield. In the reaction of allyl o-tolyl ether, ionic liquid and Lewis acid were recovered more than >95% yield after extracting the product with diethyl ether, and it is a very convenient Lewis acid as a reusable catalyst (entries 7 and 8) .[16]

In the case of using 2-methyl-2-propenyl phenyl ether as a starting material, 2,3-diisopropylbenzo[b]furan was obtained in 15% yield. So far the reaction mechanism is still not clear, however, it seemed that the claisen rearrangement occurred at first to produce the intermediate.

a Sc(OTf)$_3$ (5 mol%), 200 °C, ionic liquid, 10 h

Table 7. Lewis acid catalyzed sequential synthesis

Entry	Substrate	ionic liquid[a)	Yield (%)
1	allyl phenyl ether	1	62
2	allyl p-tolyl ether	1	88
3		2	40
4		3	12
5		4	9
6	allyl o-tolyl ether	1	91
7		1	95 [b]
8		1	90 [c]

[a] **1** [EtDBu][OTf]; **2** [MeDBU][OTf]; **3** [bmim][BF$_4$]; **4** [bmim][PF$_6$]
[b] second cycle [c] third cycle

Horner-Wadsworth-Emmons reaction

Usually, fluorinated alkenes are prepared from the addition-elimination of polyfluoroethenes, chlorofluorocarbene, ethyl phenylsulfinylfluoroacetate, triethyl 2-fluoro-2-phosphonoacetate and organometallics in polar organic solvents such as DMF, DMSO, THF etc., and after quenching with water, the products are extracted with organic solvents.[17] These processes generate the waste containing solvent media and water. Therefore, we examined the preparation of α-fluoro-α,β-unsaturated esters by the Horner-Wadsworth-Emmons reaction.[18]

PhCHO + (EtO)$_2$P(O)CHFCO$_2$Et

Z-isomer   E-isomer

a [EtDBU][OTf], base, rt, 3 hr

Table 8. Preparation of olefins

	Base	Yield (%)	Recovered ionic liquid (%)	E/Z ratio
cycle 1 :	K$_2$CO$_3$	74	98	73 : 27
cycle 2 :	K$_2$CO$_3$	71	97	74 : 26
cycle 1 :	DBU	81	98	35 : 65
cycle 2 :	DBU	88	96	32 : 68

While the synthesis of α-fluoro-α,β-unsaturated esters from the reaction of triethyl 2-fluoro-2-phosphonoacetate with aldehydes was reported by several groups,[17] their procedures required NaH as a base at diethyl ether reflux temperature or n-BuLi as a base at -78°C in tetrahydrofuran. The synthesis of α-fluoro-α,β-unsaturated esters was established from the reaction of triethyl 2-fluoro-2-phosphonoacetate with aldehydes in the presence of K$_2$CO$_3$ or 1,8-diazabicyclo-[5,4,0]-7-undecene (DBU) as a base at room temperature in the reusable ionic liquid (Table 8). The present reaction summarized in Table 9 smoothly proceeded at room temperature in an ionic liquid. The produced α-fluoro-α,β-unsaturated ester was separated by the extraction with diethyl ether, and ionic liquid was recovered more than >95%.

a ionic liquid, base, rt E-isomer Z-isomer

Table 9. Preparation of α-fluoro-α,β-unsaturated esters

Entry	RCHO R	Ionic liquid [a]	Base	Yield (%) [b]	E/Z ratio	Entry	RCHO R	Ionic liquid [a]	Base	Yield (%) [b]	E/Z ratio
1	Ph	1	K$_2$CO$_3$	74	73 : 27	6	Ph	3	DBU	79	35 : 65
2		2	K$_2$CO$_3$	68	69 : 31	7	C$_6$H$_{13}$	1	K$_2$CO$_3$	35	67 : 33
3		3	K$_2$CO$_3$	65	70 : 30	8	C$_6$H$_{13}$	1	DBU	77	33 : 67
4		1	DBU	81	35 : 65	9	PhCH(CH$_3$)	1	K$_2$CO$_3$	37	71 : 29
5		2	DBU	70	39 : 61	10	PhCH(CH$_3$)	1	DBU	69	30 : 70

a ionic liquid: **1** [EtDBU][OTf]; **2** [MeDBU][OTf]; **3** [emim][OTf] b Yield was determined by ^{19}F NMR integral intensity using benzotrifluoride as a internal standard.

Synthesis and Reaction of Zinc Reagents in Ionic Liquids

The generation of the organometallic reagents such as the Reformatsky-type reagents[19] based on the use of α-halo ester and carbonyl substrate to a suspension of zinc powder in an ionic liquid, is the next synthetic application. When the reaction was attempted in tetrahydrofuran and in ionic liquid ([EtDBU][OTf]) at room temperature, the yields of Reformatsky reaction are 65 and 52%, reapectively. However, under heating at 50-60 °C the reaction has been smoothly proceeded to give the target material in 76% yield. Table 10 summarized the results of Reformatsky-type reactions. The reactions of ethyl bromodifluoroacetate with benzaldehyde proceeded in an ionic liquid ([EtDBU][OTf])-zinc system at 50-60 °C to give the corresponding carbinol. Furthermore, in this system, as reaction intermediates (zinc reaction intermediates) was converted to the carbinols with small amount of water containing diethyl ether, ionic liquids were recovered smoothly after extracting the target material with diethyl ether. Moreover, in this reaction system, the yield was increased to 93% based on the change of molar ratio of PhCHO : $BrCF_2CO_2Et$ (1 : 3 ; entry 5). Successive reuse of the recovered ionic liquid ([EtDBU][OTf], entries 3 and 4) and in the same reaction yielded amounts of product as high as in the first cycle shown in Table 10. In the third cycle, reuse of ionic liquids recovered from the second cycle is possible to produce the same alcohol in the same reaction.[20]

^{a} ionic liquid, Zn

Table 10. Reformatsky Reaction

Entry	$BrCX_2COY$	Ionic Liquid$^{a)}$	Yield (%)	Entry	$BrCX_2COY$	Ionic Liquid$^{a)}$	Yield (%)
1	$BrCF_2CO_2Et$	1$^{b)}$	52	5	$BrCF_2CO_2Et$	1	93$^{e)}$
2		1	76	6	$BrCH_2CO_2Et$	2	61
3		1	89$^{c)}$	7		3	69
4		1	56$^{d)}$	8		1	63

^{a} **1** [EtDBU][OTf]; **2** [bmim][PF$_6$]; **3** [bmim][BF$_4$] ^{b} Reaction temperature: rt. Other entries were carried out at 50-60 °C. ^{c} second cycle ^{d} third cycle
^{e} The molar ratio of PhCHO : Zn : $BrCF_2CO_2Et$ = 1 : 3: 3.

Recently, the preparation of propargylic alcohols by direct addition of teminal alkynes to aldehydes yield in the system Et_3N-toluene at 23°C, have been reported that the yields of these reactions were in 50 ~ 99%.[21] To clarify the generation and/or synthetic scope of alkynyl zinc reagents in ionic liquids, we examined the direct addition of terminal alkynes to aldehydes in the presence of zinc trifluoromethansulfonate $(Zn(OTf)_2)$ and base. The reactions proceeded smoothly under the conditions (terminal alkyne (3 eq.), aldehyde (1.5 eq.), $Zn(OTf)_2$ (2 eq.), and 1,8-diazabicyclo[5,4,0]-7-undecene (DBU, 3 eq.) in an ionic liquid ([EtDBU][OTf]) at room temperature), giving the corresponding propargylic alcohols. The reaction summarized in Table 11 smoothly proceeded at room temperature in an ionic liquid.

In this reaction system, no reaction proceeded in the absence of base and/or $Zn(OTf)_2$. Therefore, initially the alkynylzinc reagents were prepared *in situ* from the reaction of terminal alkynes and $Zn(OTf)_2$ in the presence of base in an ionic liquid, and then the reaction of zinc reagents with aldehydes was proceeded. After 24-48 h of stirring at room temperature, the cerresponding propargyl alcohol was obtained by the extraction with diethyl ether, and ionic liquid was recovered.

$$RCHO \ + \ R'C \equiv CH \xrightarrow{\text{a}} R\text{—}\underset{R'}{\overset{OH}{\diagdown}}$$

$^a Zn(OTf)_2$, DBU, ionic liquid

Table 11. Preparation of Propargylic Alcohols

Entry	RCHO R	alkyne R_I	Ionic Liquid[a]	Yield (%)	Entry	RCHO R	alkyne R_I	Ionic Liquid[a]	Yield (%)
10	Ph	Ph	1	47	15	Ph	C_6H_{13}	1	75
11			2	59	16			2	62
12			3	35	17	$4\text{-}FC_6H_4$	Ph	1	47
13		C_4H_9	1	51	18		C_4H_9	1	50
14			2	60	19		C_6H_{13}	1	55

^{a} **1** [EtDBU][OTf]; **2** [bmim][BF$_4$]; **3** [bmim][PF$_6$]

References

1.   (a) Gladysz, J. A. *Science*, **1994**, *266*, 55. (b) Zhu, D.-W. *Synthesis*, **1993**, 953. (c) Studer, A.; Jeger, P.; Wipf, P.; Curran, D. P. *J. Org. Chem.*, **1997**,

62, 2917. (d) Horváth, I. T. *Acc. Chem. Res.* **1998**, *31*, 641 and references cited therein. (e) Hoshino, M.; Degenkolb, P.; Curran, D. P. *J. Org. Chem.*, **1997**, *62*, 8341.

2. Kobayashi, S.; Wakabayashi, T.; Nagayama, S.; Oyamada, H. *Tetrahedron Lett.* **1997**, *38*, 4559 and references cited therein.

3. (a) Earle, M. J.; McCormac, P. B.; Seddon, K. R. *J. Chem. Soc. Chem. Commun.* **1998**, 2245. (b) Adams, C. L; Earle, M. J.; Roberts, G.; Seddon, K. R. *J. Chem. Soc. Chem. Commun.* **1998**, 2097. (c) Ellis, B.; Keim, W.; Wasserscheid, P. *J. Chem. Soc. Chem. Commun.* **1999**, 337.

4. (a) Halida, S.; Curran, D. P. *J. Am. Chem. Soc.* **1996**, *118*, 2531. (b) Juliette, J. J. J.; Rutherdord, D.; Horváth, I. T.; Gladysz, J. A. *J. Am. Chem. Soc.* **1999**, *121*, 2696 and references cited therein.

5. (a) Horváth, I. T.; Rábai, J. *Science*, **1994**, *266*, 72. (b) DiMagno, S. G.; Dussault, P. H.; Schultz, J. A. *J. Am. Chem. Soc.*, **1996**, *118*, 5312. (c) Klement, I.; Lütjens, H.; Knochel, P. *Angew. Chem., Int. Ed. Engl.*, **1997**, *36*, 1454.

6. For reviews, see: (a) Hosomi, A. *Acc. Chem. Res.*, **1988**, *21*, 200. (b) Fleming, I.; Dunogues, J.; Smithers, R. *Org. React.*, **1989**, *37*, 57.

7. Banks, R. E.; Smart, B. E.; Tatlow, J. C. *Organofluorine Chemistry*; Principle and Commercial Applications, Plenum Press, New York, 1994.

8. Nakano, H.; Kitazume, T. *Green Chemistry* **1999**, *1*, 21.

9. Olah, G. A. *Friedel-Crafts Chemistry* Wiley, New York, 1973.

10. Ogawa, A.; Curran, D. P. *J. Org. Chem.*, **1997**, *62*, 450.

11. Trost, B. M. *Angew. Chem. Int. Ed. Engl.*, **1995**, *34*, 259.

12. (a) Legler, G. *Adv. Carbohydr. Chem. Biochem.* **1990**, *48*, 319. (b) Kitazume, T.; Murata, K.; Okabe, A.; Takahashi, Y.; Yamazaki, T. *Tetrahedron: Asymmetry* **1994**, *5*, 1029.

13. Zulfiqar, F.; Kitazume, T. *Green Chemistry*, **2000**, *4*, 137.

14. Kitazume, T.; Zulfiqar, F.; Tanaka, G. *Green Chemistry*, **2000**, *4*, 133.

15. Trost, B. M.; Fleming, I.; Semmelhack, M. F., Eds.; *Comprehensive Organic Synthesis;* Pergamon Press: Oxford, U.K., 1991;Vol. 5, p. 827

16. Zulfiqar, F.; Kitazume, T. *Green Chemistry*, **2000**, *6*, 296.

17. (a) Kitazume, T.; Yamazaki, T. *Experimental Methods in Organic FluorineChemistry;* Kodansha & Gordon and Breach Science: Tokyo, 1998. (b) Hudlicky, H.; Pavlath, A. E. *Chemistry of Organic Fluorine Compounds II: a Critical Review;* American Chemical Society: Washington, DC, 1995.

18. Kitazume, T.; Tanaka, G. *J. Fluorine Chem.* **2000**, *106*, 211,

19. Rathke, M. W.; Weipert, P. *Comprehensive Organic Synthesis,* Vol. 2, p. 277, Pergamon Press, Oxford, 1991.

20. Kitazume, T.; Kasai, K. *Green Chemistry*, **2001**, *3*, 30.

21. Frantz, D. E.; Fässler, R.; Carreira, E. M. *J. Am. Chem. Soc.* **2000**, *122*, 1806.

Chapter 6

Enzyme Activity Using a Perfluoropolyether-Modified NAD(H) in Fluorous Solvents and Carbon Dioxide

Janice L. Panza[1,2], Alan J. Russell[1,2], and Eric J. Beckman[1,2]

[1]Department of Chemical and Petroleum Engineering, University of Pittsburgh, 1249 Benedum Hall, Pittsburgh, PA 15260
[2]Center for Biotechnology and Bioengineering, University of Pittsburgh, 300 Technology Drive, Pittsburgh, PA 15219

The purpose of this research was to investigate the viability of developing biocatalysis in "green" solvents such as fluorous solvents and CO_2 using a soluble cofactor. The cofactor nicatinamide adenine dinucleotide (NAD) was covalently modified with a perfluoropolyether. Solubility studies of the modified NAD (FNAD) were performed in CO_2 by generating cloud point curves. At weight percents 1.2% to 2%, FNAD demonstrated solubility in liquid CO_2 at pressure above 1300 psi. The activity of horse liver alcohol dehydrogenase (HLADH) using FNAD as the coenzyme was investigated in the fluorous solvent methoxy-nonafluorobutane (HFE). The activity increased with an increase in the amount of FNAD added. The activity using FNAD was compared with the activity using the same molar amount of unmodified NAD. The activity using FNAD was greater than that of NAD. Preliminary activity studies in CO_2 demonstrated that FNAD was functional; however, future kinetic studies are necessary.

Enzymes are biological catalysts. They are proteins, which are polymers of amino acids that fold in unique three-dimensional structures forming an active site on the surface of the folded molecule. They increase the reaction rates 10^6-10^{14} over uncatalyzed reactions, and several orders of magnitude over the correspondingly chemically catalyzed reactions. Enzymes are active under mild conditions such as temperatures less than $100°C$, atmospheric pressure, and neutral pHs; whereas chemically catalyzed reactions usually require extreme temperatures, pressures, and pHs. Finally, enzymes are very specific and selective in the reactions that they catalyze. They exhibit regioselectivity, stereoselectivity, and chemioselectivity (*1*).

The characteristics of enzymes just mentioned are for enzymes in their natural environment, water. Many factors exist that make using enzymes in non-aqueous media more advantageous than in aqueous media. First, many substrates are hydrophobic making biocatalysis not possible in water (*2*). By performing enzyme-catalyzed reaction in organic media, many new substrates can be used which broadens the scope of enzyme catalysis. In addition, the use of these substrates may lead to novel reactions in organic solvents, increased specificity of an enzyme for a substrate, and the elimination of unwanted side reactions such as hydrolysis (*2,3*). Product recovery from organic solvents would be easier. Enzymes are also less sensitive to temperature in organic solutions (*3*). A final benefit to biocatalysis in organic solvents is reduced microbial contamination in the bioreactors (*2,3*).

Taken a step further, carbon dioxide can be used as a solvent for biocatalysis. CO_2 offers the same advantages of organic solvents but also has many more. CO_2 is abundant, inexpensive, nonflammable, and nontoxic. It is proposed as a "green" alternative to traditional organic solvents in that it is not regulated as a volatile organic chemical (VOC) nor restricted in food or pharmaceutical applications. Under supercritical conditions ($T_c=31°C$ and $P_c=73$ bar), CO_2, like all supercritical fluids, offers many mass transfer advantages over conventional organic solvents due to their gas-like diffusivities, low viscosities, and low surface tensions.

Fluorous solvents are also considered somewhat environmentally benign. The have low toxicity, do not deplete the ozone layer, and they are easy to separate from both aqueous and organic solvents because they are immiscible in both.

Nicatinamide Adenine Dinucleotide in Biocatalysis

For some enzymes, all that is required is for activity is binding of the substrate to the active site of the enzyme. Other enzymes require the participation of an additional molecule called a cofactor. A cofactor can either be a simple inorganic ion, such us Fe, Mg, or Zn, or it can be a complex organic or metalloorganic molecule called a coenzyme. Some enzymes require both a coenzyme and one or more metal ions for activity. Cofactors may be necessary to just assist in the binding of the substrate to the active site of the enzyme or they may be required to take part in the reaction; for example, some reactions require the transfer of a specific functional group from the substrate to the cofactors. When a cofactor is covalently bound to the enzyme, it is called a prosthetic group. A complete, catalytically active enzyme together with its cofactor is called a holoenzyme. Figure 1 demonstrates how a cofactor and enzyme work together.

One class of enzymes that require a cofactor is an oxidoreductase. These enzymes are commonly known as dehydrogenases. They catalyze oxidation/reduction reactions. Dehydrogenases require the coenzyme nicatinamide adenine dinucleotide (NAD) to participate in the reaction. NAD accepts a hydride ion from the substrate allowing the substrate to be oxidized and the NAD to be reduced to NADH. The general reaction scheme is shown where AH_2 is the reduced form of the substrate and A is the oxidized product.

$$NAD + AH_2 \rightleftharpoons NADH + A + H^+$$

The chemical structure of NAD is shown in Figure 2. NAD is composed of two nucleotides, nicatinamide mononucleotide joined to adenine mononucleotide by a phosphodiester linkage. Reduction takes place on the nicatinamide group. NAD/NADH is referred to as NAD(H).

Regeneration of NAD(H)

Problems arise when coenzymes such as NAD(H) are used to catalyze reactions outside of their native environment, the cell. The major problem is the high cost of both the enzyme and the cofactor, especially the cofactor since it is needed in stoichiometric amounts. The price of expensive cofactors such as NAD(H) has prevented the development of large-scale systems. *In situ* methods of regeneration are necessary to make the systems more economically feasible. Common regenerative strategies can be classified into 4 categories: enzymatic, electrochemical, chemical/photochemical, and biological (*4*). Each method has both advantages and disadvantages so the regenerative strategy chosen should be the one most compatible with the system.

Figure 3 demonstrates the *in situ* enzymatic recycling of NAD(H). Once NADH has been oxidized to NAD, it can be regenerated back to NADH by an enzymatic method. One enzyme, E1, oxidizes NADH to NAD and a second enzyme, E2, reduces NAD back to its original state, NADH. At the same time,

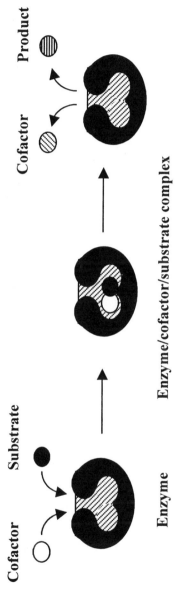

Figure 1. Schematic of how a coenzyme and enzyme catalyze a reaction.

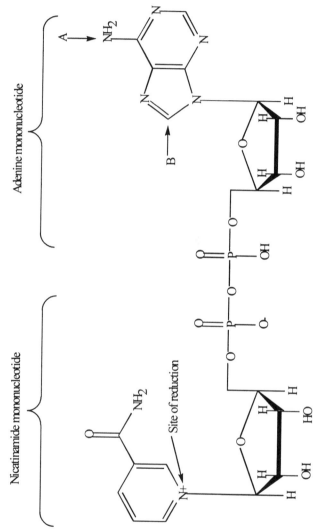

Figure 2. Structure of nicatinamide adenine dinucleotide (NAD).

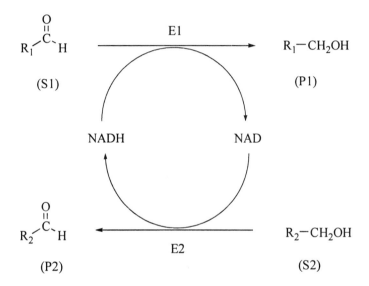

Figure 3. Recycling of NAD/NADH.

E1 reduces a substrate, S1, to a product, P1, and E2 oxidizes a substrate, S2, to product, P2. Sometimes only one enzyme is required to carry out both reductions and oxidations, necessitating the enzyme has a broad specificity.

Modification of NAD

In addition to cofactor regeneration, immobilization techniques have also been used in order to make the reactions more economically feasible. NAD has been covalently modified at position N-6 (5) and C-8 (6) (A and B, respectively in Figure 2) on the adenine group in order to be attached to solid supports.

Covalently modified NAD has several applications. First, immobilized NAD was attached to solid supports for use in affinity chromatography techniques for the purification of enzymes (6,7). Second, NAD was covalently modified with polymer to increase its molecular weight for continued reuse in an enzyme reactor (8,9). Finally, it was directly tethered to enzymes for use in enzyme reactions (10,11). In each case, even though NAD had been covalently modified, NAD retained its functionality.

Objective and Specific Aims

When NAD is used with an enzyme in an organic solvent, they are both insoluble. The enzyme and NAD are therefore initially lyophilized together; however, they are surrounded by what is termed "essential water", or the bound water that is required for the enzyme to have activity in the organic solvent (12). Once NAD participates in a reaction, it is reduced to NADH (or vice versa). It is hypothesized that NADH must be released from the active site in order for it to be regenerated (13). Since NADH is not soluble in the organic solvent, NADH does not want to leave the active site. If NADH is hindered from leaving the active site, the enzyme molecule will become inactive given that the active site is not free to bind another NAD molecule to continue the reaction. Unless a single enzyme, two-substrate recycling is happening, Deetz and Rozzell demonstrated that this was the case (14). NAD(H) would not traverse the organic phase to participate in both reactions (Figure 3) if two enzymes (E1 and E2) were lyophilized separately; the insolubility of the cofactor in the organic solvent prevented the regeneration of the cofactor. However, they showed that both reactions occurred when the two enzymes and NAD were lyophilized together. The reason for this was that the "essential water" bound to the lyophilized solid containing both enzymes offers a layer of hydration so that the cofactor can be released from the active site of E1 to migrate to the active site of E2 through the hydration layer. When the enzymes are lyophilized separately, the two enzymes do not share a hydration layer; therefore, NAD cannot travel from E1 to E2 through the organic solvent. We propose to overcome this limitation by solubilizing the NAD into the solvent.

Figure 4 demonstrates the purpose of this work. In the current situation, both the enzyme and the NAD(H) are insoluble in a fluorous and CO_2 phase since they are both hydrophilic molecules. After a reaction, the NAD(H) would be bound to the active site, but it would need to be released in order to be regenerated to participate in another reaction. Our objective to is covalently modify NAD(H) with molecules that enhance the solubility of NAD(H) into the fluorous or CO_2 phase thereby facilitating the release of the NAD(H) into the fluorous or CO_2 phase so that it can be regenerated.

In order to satisfy the objective of this project, three specific aims must be met:

1. Identify synthetic methods by which a coenzyme can be covalently modified with compounds in order to enhance solubility in fluorous solvents and CO_2.
2. Measure the effect of modification on solubility of the coenzyme in fluorous solvents and CO_2.
3. Determine the biological activity of the modified coenzyme in fluorous solvents and CO_2.

Synthesis of Perfluoropolyether-modified NAD(H)

The main problem of CO_2 is that it is a poor solvent for most compounds, so identifying a molecule with which to modify NAD is a challenge. Generally, molecules are thought of as hydrophilic or lipophilic. However, neither of these types of molecules demonstrate appreciable solubility in CO_2. Recently, compounds have be classified as CO_2-philic (*15*) or CO_2-phobic, with most hydrophilic and lipophilic molecules being classified as CO_2-philic. Fluorinated compounds such as fluoroethers and fluoroacrylates have been established as CO_2-philic. They are the compounds of interest in which to modify NAD since NAD modified with a fluorinated polymer should demonstrate solubility in CO_2 as well as fluorous solvents. For our purposes, a perfluoropolyether (PFPE) with a molecular weight of 2500 was used as the fluorinated polymer.

NAD was modified with a fluorinated polymer in a four-step reaction process as follows (Figure 5):

1. The first step is the bromination of the NAD at the C8 position of the adenine group to form NAD-Br (*6*).
2. The second step is the chemical reduction of the NAD-Br to NADH-Br using sodium dithionite (*16*). NAD is chemically reduced because the reduced form is more thermally stable than the oxidized form.

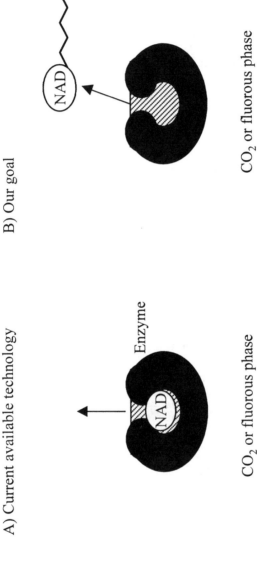

A) Current available technology

B) Our goal

Enzyme

CO_2 or fluorous phase

CO_2 or fluorous phase

Figure 4. A) Enzyme and NAD are both insoluble in the CO_2 or fluorous phase.
B) Modified-NAD to enhance solubility in CO_2 or fluorous phase

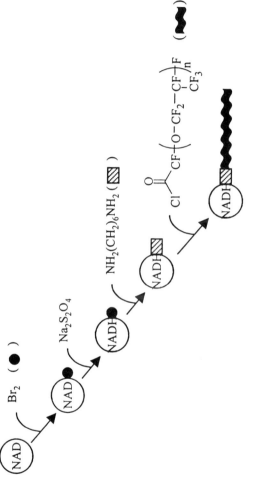

Figure 5. Synthesis of FNAD.

3. In the third reaction, the bromide group is replaced by hexanediamine (HDA) in a nucleophilic substitution reaction to form NADH-HDA. One of the free amine groups replaces the bromide at the C8 position on adenine, but the second amine group remains free for the last reaction.

4. In the final reaction, PFPE acid chloride molecule is attached to the free amino group attaching PFPE to NAD by an amide linkage. The fluorinated NAD is called FNAD.

FNAD demonstrated solubility in such fluorinated solvents as freon, perfluorodimethylcyclohexane, perfluorohexane, perfluoroheptane, and methoxy-nonafluorobutane.

Solubility Studies in CO_2

Materials and methods

Solubility studies were performed at room temperature (21°C) in a variable volume high-pressure view cell shown in Figure 6. A known amount of sample and CO_2 were added to the view cell in order to get a specific weight percent of FNAD in CO_2. The pressure of the CO_2 in the system was increased using the gas booster and syringe pump 1 until the FNAD in the CO_2 existed as one phase (FNAD was soluble in CO_2). The volume of the cell was then expanded using syringe pump 2 to decrease the pressure until the solution inside the cell became turbid. At that pressure, the FNAD was no longer soluble in the CO_2 and it was labeled the cloud point pressure. Any pressure below the cloud point was then two phases. More CO_2 was then added to the system to dilute the sample and the cloud point pressure was determined for the new weight percent of FNAD in CO_2. A cloud point curve was generated by determining the cloud point pressures at different weight percents of FNAD in CO_2.

Results

Figure 7 shows the cloud point curve of FNAD in liquid CO_2 at different weight percents. The range of weight percents is about 1.2 to 2 wt%, which corresponds to approximately 3 to 5 mM, the concentration we would expect to use in a reaction. As can be seen by the graph, the cloud point pressures are relatively constant over the weight percent range of interest, the cloud point being between 1200 to 1400 psi. At pressure above the cloud point curve, the system will be one phase. At pressures below the cloud point curve, the system will be two phase, since the FNAD will not be soluble in the liquid CO_2 at those pressures. Since the goal is to perform enzyme catalyzed oxidation/reduction

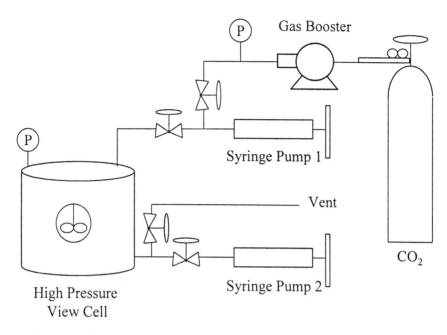

Figure 6. Diagram of the high pressure view cell to study solubility in CO_2.

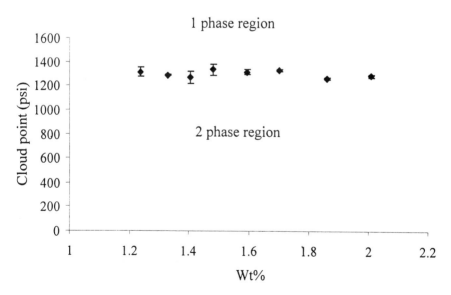

Figure 7. Cloud point curve of FNAD in liquid CO_2.

reactions in CO_2 with a CO_2 soluble cofactor, we have to use a pressure where the FNAD exists as one phase with the CO_2, or at a pressure above the cloud point curve.

Activity in a Fluorous Solvent

The next aim was to determine the biological activity of the modified NAD in fluorous solvents and CO_2. Since the enzymes and cofactors were altered and used in unnatural environments, it was necessary show that they still function. Activity studies in a fluorous solvent were performed in methoxy-nonafluorobutane (3M). It is called HFE since it is a hydrofluoroether. HFE is intended to replace other ozone-depleting solvents such as CFCs. This fluid's benefits include zero depletion potential, very low toxicity, and non-flammability.

One enzyme, horse liver alcohol dehydrogenase (HLADH), was used to catalyze both reactions instead of two enzymes as shown in Figure 3. HLADH demonstrates broad substrate specificity, so it can be used to catalyze both reactions in the activity studies.

Materials and Methods

Different amounts of FNAD were added to 5 mg HLADH and 250 mM butyraldehyde and ethanol (S1 and S2 respectively in Figure 3) in HFE. The total reaction volume was 2 ml. The reaction was shaken at 250 rpm and 30°C. The reaction mixtures were analyzed for the product, butanol (P1), using a FID-GC. The second product (P2) was acetaldehyde. Acetaldehyde is very volatile (bp 21°C) and was expected to vaporize, since the reactions were run at 30°C. Care was taken to open the vials in the hood before sampling.

Results

Figure 8 shows the results of the activity studies. The production of butanol in mM is plotted on the y-axis and the time in minutes on the x-axis. Each sample contained increasing amounts of FNAD from approximately 0 to 0.2 g (0 to 30 mM). As you can see from the graph, increasing the amount of FNAD increases the rate of reaction; however, the activity of the 0.2 g sample demonstrates less activity than both the 0.167 g and 0.114 g samples. It is possible that there is an optimum amount of FNAD required and more than optimum may not increase the activity. Yang and Russell demonstrated that there is an optimum molar ratio of NAD to enzyme and adding more NAD does not increase the activity (*13*).

The sample that does not contain any added FNAD (just enzyme and substrate) showed activity; therefore, the enzyme must have residual bound NAD present. The reason for this is that when HLADH is purified, NAD is added in order to stabilize the enzymes during the purification step.

Since the enzyme alone shows activity in HFE, the question to ask is whether unmodified NAD (insoluble NAD) shows the same activity as the soluble FNAD. In order to address this question, the activity of NAD was compared with the same molar amount of FNAD using the same reaction scheme. Initially, HLADH (10 mg) was added to 250 mM of both butyraldehyde and ethanol in HFE (4 ml). The reaction proceeded for 6 hours at which point, the reaction was split into two samples. In one sample, insoluble NAD was added. To the other sample, the soluble FNAD was added at the same molar concentration (9mM). The reactions were allowed to proceed monitoring the production of butanol. Figure 9 shows the results. As is evident by the graph, soluble FNAD had a faster reaction rate than insoluble NAD.

Activity in CO_2

Materials and methods

Figure 10 shows the schematic of the high pressure reactor used to study the activity in CO_2. The substrates and enzyme were added to the reactor, followed by the addition of CO_2 at room temperature (21°C) and 2500 psi (well above the cloud points of the substrates and FNAD). The mixture was stirred and the reaction mixture was bubbled through a solvent, heptane, through a 200 μl sample port in order to monitor the production of butanol.

Results

First, it was necessary to determine if all of the substrates were soluble in liquid CO_2 at the required concentrations (250 mM). Both ethanol and acetaldehyde are miscible with CO_2 under these conditions. The cloud points of butanol and butyraldehyde were generated, as shown in Figure 11, using the high-pressure view cell (Figure 6). So at a concentration of 250 mM, which is the substrate concentration that we used when the reactions were performed in the HFE, the reactions must be performed at a pressure greater than 1800 psi to ensure that all components of the reaction are soluble in CO_2 (except for the enzyme). Recall that FNAD was soluble in CO_2 above approximately 1300 psi.

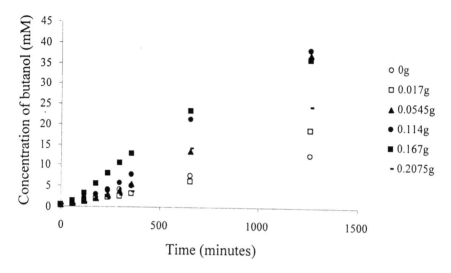

Figure 8. Activity of HLADH using FNAD in HFE.

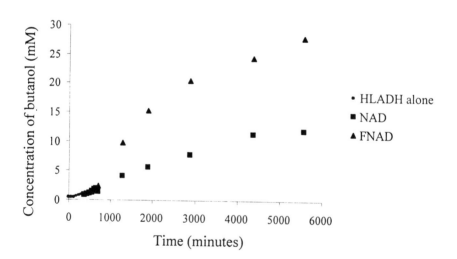

Figure 9. Coenzyme activity of FNAD (soluble) vs. NAD (insoluble) in HFE.

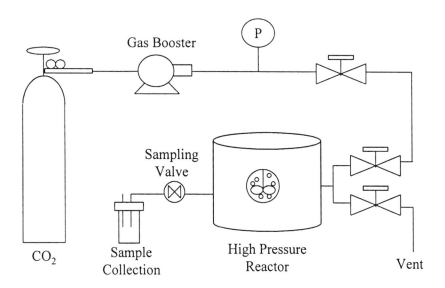

Figure 10. Schematic of high pressure reactor.

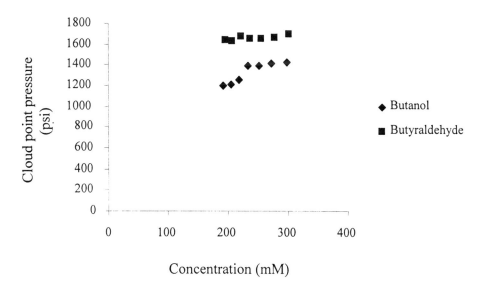

Concentration (mM)

Figure 11. Cloud points curves of butyraldehyde and butanol in liquid CO_2.

Preliminary results showed that when the substrates, butyraldehyde and ethanol, along with HLADH and the FNAD were added to the reactor, butanol was produced with time. However, future studies will investigate the kinetics in CO_2 under these conditions.

Conclusions

We have synthesized NAD with an attached perfluoropolyether, called FNAD. The modified NAD demonstrated solubility in several fluorinated solvents. In addition, we have generated cloud point curves that demonstrate that 1.2 to 2 wt % of FNAD is also soluble in CO_2 at pressure above about 1300 psi. Finally, we have showed that FNAD retains its functionality and is able to catalyze oxidation/reduction reactions along with the enzyme HLADH in the fluorous solvent HFE. In addition, preliminary results suggest that FNAD is active in CO_2.

References

1. Margolin, A. L. CHEMTECH **1991,** *March*, 160-167.
2. Chatterjee, S.; Bambot, S.; Bormett, R.; Asher, S.; Russell, A.J. In *Encyclopedia of Molecular Biology*; Kendrew J., Ed.; Oxford: Cambridge, MA, 1994; 249-258.
3. Klibanov, A. M. CHEMTECH **1986,** *June*, 354-359.
4. Chenault, H.K.; Whitesides, G.M. Applied Biochemistry and Biotechnology **1987,** *14*, 147-195.
5. Vanhommerig, S.A.M.; Sluyterman, L.A.Æ.; Meijer, E.M. Biochimica et Biophysica Acta **1996,** *1295*, 125-138.
6. Lee, C.Y.; Kaplan, N.O. Archives of Biochemistry and Biophysics **1975,** *168*, 665-676.
7. Lee, C.Y.; Lappi, D.A.; Wermuth, B.; Everse, J.; Kaplan, N.O. Archives of Biochemistry and Biophysics **1974,** *163*, 561-569.
8. Bückmann, A.F.; Kula, M.R.; Wichmann, R.; Wandrey, C. J. Appl. Biochem. **1981,** *3*, 301-315.
9. Wichmann, R.; Wandrey, C.; Bückmann, A.F.; Kula, M.R. Biotechnol. Bioeng. **1981,** *23*, 2789-2802.
10. Nakamura, A.; Urabe, I.; Okada, H. Journal of Biological Chemistry **1986,** *261*, 16792-16794.
11. Eguchi, T.; Iizuka, T.; Kagotani, T.; Lee, J.H.; Urabe, I; Okada, H. Eur. J. Biochem. **1986,** *155*, 415-421.
12. Schulze, B.; Klibanov, A.M. Biotechnol. Bioeng. **1991,** *38*, 1001-1006.
13. Yang, F.; Russell, A.J. Biotechnol. Prog. **1993,** *9*, 234-241.

14. Deetz, J.S.; Rozzell, J.D. Annals New York Academy of Sciencs **1988**, *542*, 230-234.
15. DeSimone, J.M.; Maury, E.E.; Menceloglu, Y.Z.; McClain, J.B.; Romack, T.J.; Combes, J.R. Science **1994,** *265,* 356-359.
16. Lehninger, A.L. Methods Enzymol. **1957,** *3*, 885-887.

Chapter 7

Separation of Species from Ionic Liquids

Joan F. Brennecke, Lynnette A. Blanchard[1], Jennifer L. Anthony,
Zhiyong Gu[2], Isidro Zarraga[3], and David T. Leighton

Department of Chemical Engineering, University of Notre Dame,
Notre Dame, IN 46556
[1]Current address: Intel Massachusetts, HD1–471, 75 Reed Road,
Hudson, MA 01749–2895
[2]Current address: Department of Chemical Engineering, State University
of New York at Buffalo, Buffalo, NY 14260–4200
[3]Current address: School of Chemical Engineering, Georgia Institute of Technology,
778 Atlantic Drive, Atlanta, GA 30332

The tremendous interest in the use of ionic liquids as
environmentally benign solvents for reactions presents the
need to develop equally environmentally friendly methods to
separate target solutes from mixtures containing ionic liquids.
Here we discuss various separation methods and present
results for phase behavior and physical properties important
for separations. In particular, we show that liquid/liquid
extraction with water will inevitably result in the loss of some
ionic liquid solvent and that a variety of solutes can be
extracted from ionic liquids with supercritical CO_2 without
cross-contamination. In addition, we show that pure ionic
liquids are very incompressible and do not show any evidence
of shear thinning.

Ionic liquids (ILs) appear to be excellent solvents for a wide variety of reactions. Some of these have been reviewed recently by Welton (1). While some reactions occur in ILs with greater rates and selectivities than in conventional solvents (2), the main interest in ILs stems from their vanishing small vapor pressures (3,4). Negligible evaporation means that IL solvents would not escape into the atmosphere, which eliminates the most common method of exposures to workers and any potential contributions to atmospheric pollution. Thus, ILs may be inherently safer and more sustainable solvents than traditional organic solvents. There has been steadily growing interest in these compounds, especially since the development of water stable ILs (e.g., 1-ethyl-3-methylimidazolium BF_4 and 1-ethyl-3-methyl-imidazolium $MeCO_2$ (5))

When ILs are used as solvents for reactions, one will inevitably need to separate products and any unreacted starting materials from the mixture. While techniques (e.g., distillation, liquid/liquid extraction, absorption, stripping, crystallization, etc.) to separate species from common organic solvents are well developed, the lack of physical properties and phase behavior data for systems involving ILs makes the development of similar separation processes for ILs particularly challenging. Moreover, one would not expect conventional thermodynamic models to give adequate representations of mixtures containing ILs. Thus, thermodynamic and physical property measurements necessary for the development of separation techniques for ILs are vitally important. Until recently, very little attention has been paid to these separations. Some of these recent studies (6-12) will be discussed below.

First, we will discuss various separation techniques and show how they might be applied to the separation of target species from mixtures containing ILs. Next, we will focus on two techniques, liquid/liquid extraction with water and extraction with supercritical CO_2. Finally, we will discuss two important physical properties; viscosity and density.

Separation Techniques

Since industrial processes seldom achieve 100% reaction conversion, the effluent from a typical reactor is likely to contain residual reactants, products, and by-products, as well as the solvent. Even in the simple case of an IL with a single product, a wide variety of potential separation techniques might be considered. Below we will discuss some possibilities, including distillation, gas stripping, crystallization, liquid/liquid extraction, and supercritical fluid extraction. There is also the potential for membrane separations, adsorption, chromatography, and the use of chelating agents for metal ions, among others, but we will leave discussion of these techniques to other publications. As the various techniques are described, special attention will be paid to the potential environmental impact of each of those processes. This is important since one would not want to negate the environmental gain realized by the use of the IL with an environmentally harmful separation process.

Distillation

Since ILs are essentially nonvolatile, the simplest method to remove products from an IL would be distillation. This might be simple evaporation of the products in a flash drum or wiped film evaporator. The separation could also be performed as pervaporation, using an appropriate membrane system, where the volatile species are transported from the mixture to the low pressure side of the membrane. The negligible vapor pressure of the IL means that the separation from the IL should be highly selective; i.e., the overhead product from the distillation column, flash drum, wiped film evaporator, or pervaporation system should contain essentially no IL. However, this method is limited to volatile compounds that are not thermally labile (i.e., they do not decompose when heated). The main environmental impact of distillation or evaporation is the energy use associated with vaporizing the target species.

Gas Stripping

Another alternative for the removal of volatile components from an IL is gas stripping. This involves bubbling air or some other inert gas through the IL mixture to strip out the volatile components. The disadvantage of this technique is that the volatile products may be present in the gas stream in relatively dilute quantities and removing them from the gas by condensation may require significant refrigeration costs, which is likely to represent the major environmental impact of this separation technique. In general, one would not expect gas stripping to be particularly selective if there are multiple volatile components present. On the other hand, like distillation, there will not be any IL contamination of the product in the gas phase. The compound added to do the stripping (e.g., air) is usually nontoxic and any residual amounts of it left in the liquid product is not of any environmental or safety concern.

Crystallization

If the target species in a mixture containing an IL have melting points well below the melting point of the IL (which is frequently the case), then crystallization may be an option. For instance, in the synthesis of 1-*n*-butyl methyl imidazolium chloride ([bmim][Cl]), the [bmim][Cl] product can be removed from the reaction mixture by crystallization. However, crystallization is usually not the method of choice when the supernatant liquid is the desired product instead of the solid because the supernatant liquid is seldom pure; i.e., some of the IL will remain dissolved in the liquid. Even a simple binary system is likely to form a eutectic. Upon cooling, one can obtain an essentially pure solid but the remaining liquid will never be more pure than the eutectic composition. For a mixture of a low melting point organic and an IL, this means that some of the IL will remain in the liquid organic. In addition, refrigeration costs, and the associated environmental pollution, are frequently high for crystallization processes.

Liquid/Liquid Extraction

When target species are not volatile or are thermally labile, liquid/liquid extraction is frequently employed. This might be a natural choice for the recovery of products from IL mixtures, as well. If there are multiple organic compounds in the IL mixture that have different polarities or functionalities, then liquid/liquid extraction may afford the opportunity for some selectivity in the extraction. When considering liquid/liquid extraction there are three features of the phase behavior that are vitally important.

First, the IL and the extractant must partition into two separate phases; i.e., they must be immiscible. For instance, researchers have found that diethyl ether is immiscible with some ILs and it has been used on a laboratory scale to extract reaction products for analysis (2). 1-*n*-Butyl-3-methylimidazolium hexafluorophosphate ([bmim][PF$_6$]) is immiscible with water (6). Conversely, [bmim][BF$_4$] has been reported to be totally miscible with water at room temperature (13). However, if the temperature is lowered, then a phase split will occur (13). This is an example of a system with an upper critical solution temperature (UCST).

A second extremely important property when considering liquid/liquid extraction is mutual solubility, i.e., the solubility of the solvent in the IL-rich phase and the solubility of the IL in the solvent-rich phase. This is illustrated in Figure 1. Essentially all liquid/liquid systems, even oil and water, exhibit some, at least small, mutual solubility. For instance, the mutual solubilities of [bmim][BF$_4$] and water at 0°C are 27 wt. % water in the IL-rich phase and 25 wt. % IL in the water-rich phase (13). The solubility of the solvent in the IL-rich phase may have positive or negative consequences. On the other hand, any solubility of the IL in the solvent-rich phase is almost always a disadvantage. This solubility represents contamination of the extract with the IL and presents additional downstream separation problems. Moreover, it could result in a net loss of the IL from the system. Extreme care (and cost) would have to be exercised to recover this IL, both for economic reasons and because the environmental influence of the release of IL into the environment (for instance, in wastewater) is currently unknown.

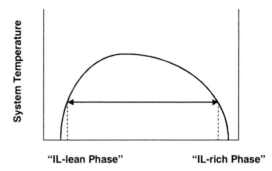

"IL-lean Phase"   "IL-rich Phase"

Figure 1   Typical liquid/liquid phase envelope for a binary mixture, showing an upper critical solution temperature (UCST).

The third important thermodynamic property of liquid/liquid extraction systems is target species partitioning, which must favor partitioning into the solvent for a viable extraction process. If the IL mixture contains multiple organic species, then the solvent could be chosen to selectively extract the desired product. While experimental measurements are preferred, partition coefficients of solutes between two phases comprised of conventional solvents can be estimated by various activity coefficient models (14). Unfortunately, these models have not yet been tested or developed for IL solutions.

In summary, liquid/liquid extraction may be an excellent way to selectively remove a target compound from a mixture containing an IL. Since many liquid/liquid extractions are performed near ambient conditions, environmental costs associated with energy usage are usually not a concern. However, there are some concerns associated with contamination of the extract phase with the IL, and subsequent downstream separation and waste treatment challenges. Paramount, though, is the choice of the extraction solvent. The use of extraction solvents that are conventional volatile organic liquid solvents might negate the environmental benefits gained from the use of the IL as the reaction solvent. Thus, requiring the liquid extractant to be environmentally benign would severely limit the choices of potential extraction solvents. Water is the only compound that clearly fits those requirements.

Supercritical CO_2 Extraction

An alternative to liquid/liquid extraction that has been used for a variety of processes, especially in the food industry, is supercritical extraction (15). Of particular interest, because it is nontoxic and nonflammable, is supercritical (SC) CO_2, whose critical temperature is just above ambient (T_c=31°C, P_c=73 bar). Although it is a greenhouse gas, CO_2 is generally considered to be an environmentally benign solvent when it is simply used in a process rather than produced. SC and liquid CO_2 are used commercially for the extraction of caffeine from coffee and tea, for the extraction of natural products like hops and saw palmetto, as a propellant for spray-painting and even as a dry cleaning solvent for individual consumers (15,16).

The advantage of supercritical extraction over liquid extraction is the tunability of the solvent that facilitates easy downstream separation. Solutes can be extracted at higher pressures, where the solvent is dense and has a high capacity for solutes. The target species can be recovered from the supercritical fluid by changing conditions so that the fluid density decreases. This can commonly be achieved by either decreasing pressure or increasing temperature. The resulting precipitation of the solute allows easy recovery of the product and recycle of the solvent. As will be shown below, extraction of solutes from ILs with supercritical CO_2 can be achieved with no measurable cross-contamination of the CO_2 with IL (11,12). The main environmental concern with supercritical extraction is the energy costs associated with compression of the CO_2.

Below we will focus on results from our laboratories associated with liquid/liquid extraction of solutes from ILs using water, and supercritical CO_2 extraction. In addition, we will present some pure component viscosity and compressibility results.

Liquid/Liquid Extraction with Water

Huddleston and coworkers (6) first proposed the separation of solutes from ILs by liquid/liquid extraction with water. In particular, they examined the partitioning of a variety of organic solutes between [bmim][PF$_6$] and water and found that these partition coefficients could be roughly correlated with 1-octanol/water partition coefficients. Subsequent work has shown that metal ions could be partitioned into an IL from an aqueous phase using a variety of extractants (7,9,10) and that ionizable organic solutes can be separated by pH-dependent reversible partitioning (8). Thus, the extraction of species from an IL with water is certainly a viable option.

For a variety of common water stable ILs, such as [bmim][PF$_6$], the first criterion, immiscibility with the extraction solvent, is satisfied for extraction with water. In addition, Rogers and coworkers (6,8-10) have shown that a favorable partition coefficient exists for a variety of species. Below we present results for the mutual solubility of several ILs with water.

As mentioned above, the mutual solubilities of the IL dissolved in the water-rich phase and the water dissolved in the IL-rich phase represent an important criterion when considering liquid/liquid extraction. We have measured the mutual solubilities with water for the following ILs:

[bmim][PF$_6$], 1-n-butyl-3-methylimidazolium hexafluorophosphate

[C$_8$-mim][PF$_6$], 1-n-octyl-3-methylimidazolium hexafluorophosphate

[C$_8$-mim][BF$_4$], 1-n-octyl-3-methylimidazolium tetrafluoroborate

[bmim][PF$_6$] was obtained from Sachem. [C$_8$-mim][PF$_6$] and [C$_8$-mim][BF$_4$] were generously given to us by the group of Prof. Seddon at the Queen's University of Belfast. All samples were used without further purification.

The mutual solubilities were determined by placing the IL in a small vial with an excess of distilled water so that only a small vapor space remained. The vials were sealed and stirred vigorously. In all cases stirring proceeded for several hours, usually overnight, with care taken so that the stirring disrupted the IL/water interface to facilitate mass transfer and ensure approach to equilibrium. Measurements were made at 22$\pm$1°C and 0.98$\pm$0.03 bar. The solubility of water in the IL-rich phase was determined by Karl-Fischer titration using a AQUASTAR V-200 Volumetric Titrator from EMScience. The solubility of the IL in the water-rich phase was determined by UV-visible absorption spectroscopy, monitoring at 211 nm. The results are shown in Table I. As expected, an increase in the length of the alkyl chain decreases the mutual solubilities with water. Conversely, replacing the PF$_6$ anion with BF$_4$ increases the mutual solubilities, with as much as 10.8 wt% water dissolving in the IL-rich phase. Also note that even for the most hydrophobic of these three compounds, the equilibrium solubility of IL dissolved in the aqueous phase is substantial (0.7 wt %). This is a serious concern if water were to be used to extract solutes from an IL because more complex separation techniques would have to be employed to remove the solute from the contaminated aqueous phase. Moreover, the IL would have to be separated from the water so that it could be recycled to the process or the aqueous phase would have to be treated before release into the environment. In this latter case, the solubility of the IL in the aqueous phase would represent a net loss of IL from the system. Note that the solubility of water in the IL-rich phase listed below for [bmim][PF$_6$] is higher than the water concentration reported by Rogers and coworkers (9). With relatively short contact times (on the order of minutes) between the aqueous and IL phases, thermodynamic equilibrium may not have been achieved. Our experience is that several hours of vigorous agitation is required to approach equilibrium, which is consistent with mass transfer between two phases of significantly different viscosities.

Table I Mutual solubilities in ionic liquid/water systems.

	IL in water-rich phase	*Water in IL-rich phase*
[bmim][PF$_6$]	2.0 wt % IL	2.3 wt % water
[C$_8$mim][PF$_6$]	0.7 wt % IL	1.3 wt % water
[C$_8$mim][BF$_4$]	1.8 wt % IL	10.8 wt % water

Extraction with Supercritical CO$_2$

As an alternative to liquid extraction we have proposed the use of supercritical CO$_2$ to remove target species from an IL. Our preliminary work (11) demonstrated the feasibility of this concept. First, we showed that [bmim][PF$_6$]/CO$_2$ is a two-phase system. This was accomplished by performing cloud point measurements, and determining the compositions of the liquid and vapor phases in equilibrium. We found that significant amounts of CO$_2$ dissolved in the IL-rich phase (up to a mole fraction of 0.6 at 80 bar) but showed that there was no measurable IL in the CO$_2$-rich phase. This is the main advantage of supercritical CO$_2$ extraction over liquid/liquid extraction. The main disadvantage, of course, is relatively high pressure operation. Also in our preliminary work (11) we showed that it was possible to extract a solute from the IL with supercritical CO$_2$. Naphthalene was chosen as a representative nonvolatile organic solute, as it dissolves readily in [bmim][PF$_6$] and CO$_2$ (17). Replicate extractions of naphthalene from [bmim][PF$_6$] with CO$_2$ at 138 bar and 40°C resulted in recoveries of 94-96%

Subsequently, we have shown that the phase behavior of [bmim][PF$_6$] with CO$_2$ is typical of a variety of IL/CO$_2$ systems (18). In particular, all of the systems are two phase, significant amounts of CO$_2$ dissolve in the IL but no measurable IL dissolves in the CO$_2$. A comparison of the solubilities of CO$_2$ in six different ILs is shown in Figure 2. In addition to the three ILs mentioned above, we studied three more samples supplied by Professor Seddon's group:

[bmim][NO₃], 1-*n*-butyl-3-methylimidazolium nitrate
[*N*-bupy][BF₄], *N*-butylpyridinium tetrafluoroborate
[emim][EtSO₄], 1-ethyl-3-methylimidazolium ethylsulfate

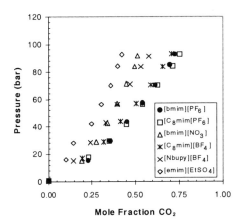

Figure 2 Liquid phase compositions of six ionic liquid - CO₂ mixtures at 40°C. (Reproduced from reference 18. Copyright 2001 American Chemical Society.)

Most recently, we have shown that it is possible to extract a wide variety of solutes from [bmim][PF₆] with supercritical CO₂ (12). These results are shown in Figures 3 and 4 for the extraction of both alkane-based solutes and aromatic solutes. The percent extracted is shown as a function of the amount of CO₂ used in the extraction. In general, solutes that have a higher solubility in CO₂ are removed more easily. Thus, very polar solid solutes like hexanoamide, that have relatively low sublimation pressures and very little affinity for CO₂, require substantial amounts of CO₂ to be removed. Nonetheless, all of the compounds can be quantitatively removed from the IL using CO₂. More importantly, this is done without any cross contamination of CO₂ with the IL.

To summarize, we have shown that CO₂ can completely separate a wide variety of organic solutes from an IL without any contamination of the extract with the IL. CO₂ dissolves in the IL liquid phase, yet the IL does not dissolve in the CO₂, allowing recovery of products in pure form. Thus, combining two solvents, both of which are generally considered environmentally benign, may offer an exciting new solution for chemical synthesis and separation problems.

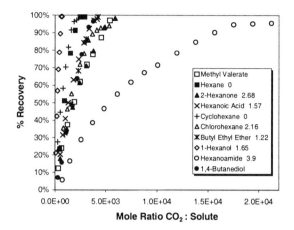

Figure 3. *The extraction of various solutes from [bmim][PF₆] with supercritical CO₂ at 138 bar and 40°C. The numbers in the legend are solute dipole moments. (Reproduced from reference 12. Copyright 2001 American Chemical Society.)*

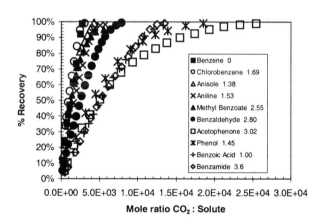

Figure 4 *The extraction of various solutes from [bmim][PF₆] with supercritical CO₂ at 138 bar and 40°C. The numbers in the legend are solute dipole moments. (Reproduced from reference 12. Copyright 2001 American Chemical Society.)*

Important Physical Properties

Basic physical properties, like melting points, boiling points, densities and viscosities, of both pure components and mixtures, are needed to design any separation process. Most researchers who have synthesized new ILs report their melting points. However, there is a paucity of other pertinent physical property data. Moreover, simple physical property measurements can provide insight into intermolecular interactions. Here we present some preliminary results for the pressure and temperature effect on the density of [bmim][PF$_6$], as well as pure component viscosity measurements for [bmim][PF$_6$]. In both cases, the [bmim][PF$_6$] is the sample obtained from Sachem, which has been dried vigorously, as described previously (12,18).

Pure Component Density Measurements

Densities of [bmim][PF$_6$] as a function of temperature were obtained in a standard glass pycnometer. Densities as a function of pressure were performed with an ultra-high-pressure apparatus that has been described elsewhere (19).

The density of [bmim][PF$_6$] as a function of temperature is shown in Figure 5. Although the density does decrease with increasing temperature, as expected, the expansion is somewhat less than is normally observed for molecular organic solvents.

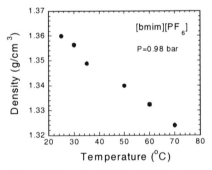

Figure 5   Density of [bmim][PF$_6$] at 0.98 bar.

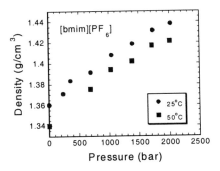

Figure 6 Density of [bmim][PF₆] as a function of pressure at 25°C and 50°C.

The density of [bmim][PF₆] as a function of pressure is shown in Figure 6 for two temperatures. Once again, the expected trend, increasing density with increasing pressure, is observed. However, [bmim][PF₆] is quite incompressible, as compared to normal molecular organic species. Its density increases about the same amount as water over the same pressure range.

Pure Component Viscosity Measurements

In separation processes, viscosity affects mixing and contacting. Viscosity measurements were performed with a Carrimed viscometer at constant shear stress, using a 6 cm diameter plate and a 0.5 mm separation. The viscosity of [bmim][PF₆] as a function of shear rate at 20°C is shown in Figure 7.

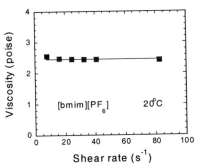

Figure 7 Viscosity of [bmim][PF₆] at 20°C as a function of shear rate.

As can be seen in the figure, ILs are quite viscous compared to normal organic solvents (typical viscosities of 0.005-0.05 poise). Moreover, this graph indicates that there is no shear-thinning. Many polymers exhibit shear-thinning; i.e., the viscosity decreases dramatically as the polymer chains line up in the direction of shear when the shear is increased. They can also exhibit nonzero normal stresses and elastic moduli. A Rheometrics Scientific instrument with a 5 cm diameter plate was used to measure the normal stresses and elastic moduli of [bmim][PF$_6$]. However, to within instrument accuracy no normal stresses were detected in the shear rate range of 1 s^{-1} to 100 s^{-1}. Furthermore, oscillatory measurements showed that the ratio between elastic and viscous moduli, G'/G'', for this material was negligible in the frequency range of 1 to 100 rad/s. Thus, [bmim][PF$_6$] appears to be a simple Newtonian fluid that shows no indication of long-range structure.

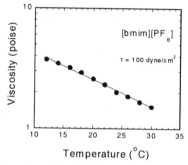

Figure 8 Viscosity of [bmim][PF$_6$] as a function of temperature.

Shown in Figure 8 is the viscosity of [bmim][PF$_6$] as a function of temperature over the relatively small range from 12°C to 30°C. As expected, the viscosity decreases with increasing temperature, by 5.2% per °C in this range . This is similar to the temperature dependence of other viscous fluids, such as silicones like Dow Corning 550.

Summary

Ionic liquids exhibit the potential to be the next generation of industrial solvents, as their total lack of volatility eliminates the most common causes of safety and environmental hazards associated with organic solvent use. We have discussed when conventional separation processes may be applicable to the recovery of target species from ILs. We have presented mutual solubility data for IL/water systems that is important in evaluating water as a solvent for liquid/liquid extraction. In addition, we have shown that supercritical CO_2 can be used to quantitatively recover solutes from ILs without any cross-contamination. Finally, we have presented density and viscosity measurements for [bmim][PF_6], which will be important in the design of any separation process using this IL solvent.

Acknowledgements

Financial support from the Environmental Protection Agency (grant R826734-01-0), the National Science Foundation (grant EE697-00537-CRCD), and Clare Booth Luce Fellowship are gratefully acknowledged. We thank Prof. M. J. McCready for his suggestions on the rheological measurements and Prof. K. Seddon at The Queen's University of Belfast for the ionic liquid samples.

References

1. Welton, T. Room-temperature ionic liquids. Solvents for synthesis and catalysis. *Chem. Rev.* **1999**, *99*(8), 2071-2084.
2. Earle, M. J.; McCormac, P. B.; Seddon, K. R. Diels-Alder reactions in ionic liquids, *Green Chemistry* **1999**, *1*, 23-25.
3. Hussey, C. L. Room temperature haloaluminate ionic liquids. Novel solvents for transition metal solution chemistry, *Pure & Appl. Chem.* **1988**, *60*(12), 1763-1772.
4. Seddon, K. R. Room-temperature ionic liquids: neoteric solvents for clean catalysis, *Kinetics and Catalysis* **1996**, *37*(5), 693-697.
5. Wilkes, J. S.; Zaworotko, M. J. Air and water stable 1-ethyl-3-methylimidazolium based ionic liquids, *J. Chem. Soc., Chem. Commun.*, **1992**, 965-967.
6. Huddleston, J. G.; Willauer, H. D.; Swatloski, R. P.; Visser, A. E.; Rogers, R. D. Room temperature ionic liquids as novel media for 'clean' liquid-liquid extraction, *J. Chem. Soc., Chem. Commun.* **1998**, 1765-1766.
7. Dai, S.; Ju, Y. H.; Barnes, C. E. Solvent extraction of strontium nitrate by a crown ether using room-temperature ionic liquids, *J. Chem. Soc., Dalton Trans.* **1999**, 1201-1202.
8. Visser, A. E.; Swatloski, R. P.; Rogers, R. D. pH-Dependent partitioning in room temperature ionic liquids, *Green Chemistry* **2000**, February, 1-4.

9. Visser, A. E.; Swatloski, R. P.; Reichert, W. M.; Griffin,S. T.; Rogers, R. D. Traditional extractants in nontraditional solvents: Group 1 and 2 extraction by crown ethers in room-temperature ionic liquids, *Ind. Eng. Chem. Res*. **2000**, *39*, 3596-3604.

10. Visser, A. E.; Swatloski, R. P.; Reichert, W. M.; Mayton, R.; Sheff, S.; Wierzbicki, A.; Davis, J. H., Jr.; Rogers, R. D. Task-specific ionic liquids for the extraction of metal ions from aqueous solutions, *J. Chem. Soc., Chem. Commun*. **2001**, 135-136.

11. Blanchard, L. A.; Hancu, D.; Beckman, E. J.; Brennecke, J. F. Green processing using ionic liquids and CO_2, *Nature* **1999**, *399*, 28-29.

12. Blanchard, L. A.; Brennecke, J. F. Recovery of organic products from ionic liquids using supercritical carbon dioxide, *Ind. Eng. Chem. Res*. **2001**, 40, 287-292.

13. Dullius, J. E. L.; Suarez, P. A. Z.; Einloft, S.; de Souza, R. F.; Dupont, J.; Fischer, J.; De Cian, A. Selective catalytic hydrodimerization of 1,3-butadience by palladium compounds dissolved in ionic liquids, *Organometallics* **1998**, *17*, 815-819.

14. Prausnitz, J. M.; Lichtenthaler, R. N.; Gomes de Azevedo, E. *Molecular Thermodynamics of Fluid-Phase Equilibria*, 3rd Ed.; Prentice Hall: Upper Saddle River, NJ, 1999.

15. McHugh, M. A.; Krukonis, V. J. *Supercritical Fluid Extraction, Principles and Practice*, 2nd Ed; Butterworth-Heinemann: Stoneham, MA, 1994.

16. Brennecke, J. F. New applications of supercritical fluids, *Chemistry and Industry* **1996**, 4 November, 831-834.

17. McHugh, M.; Paulaitis, M. E. Solid solubilities in supercritical fluids at elevated pressures, *J. Chem. Eng. Data* **1980**, *25*, 326-329.

18. Blanchard, L. A.; Gu, Z.; Brennecke, J. F. High-pressure phase behavior of ionic liquid/CO_2 systems, *J. Phys. Chem. B* **2001**, *105*, 2437-2444.

19. Gu, Z. *Experimental Studies of Thermophysical Properties of Room Temperature Ionic Liquids*; M.S. Thesis: University of Notre Dame, Notre Dame, IN, 2000.

Chapter 8

Catalysis Using Supercritical or Subcritical Inert Gases under Split-Phase Conditions

Philip G. Jessop[1], Charles A. Eckert[2], Charles L. Liotta[2], R. Jason Bonilla[1], James S. Brown[2], Richard A. Brown[1], Pamela Pollet[2], Colin A. Thomas[1], Christy Wheeler[2], and Dolores Wynne[1]

[1]Department of Chemistry, University of California, Davis, CA 95616–5295
[2]Schools of Chemistry and Chemical Engineering, Georgia Institute of Technology, Atlanta, GA 30332–0100

Compressed CO_2 and other gases can serve as media for catalysis in a number of different ways. While use of such gases as supercritical fluids (SCFs) under single-phase conditions has been studied intensively over the past decade, the use of inert gases under split phase conditions has received less attention. Having a condensed phase such as water, an ionic liquid, or even a solid below a SCF allows one to perform reactions combined with simultaneous or subsequent separation of product from catalyst or excess reagent. Performing a reaction in a condensed phase below a subcritical gas allows one to modify the reaction behaviour by adjusting the choice of inert gas or pressure of the gas. Examples of these possibilities are described, with an emphasis on applications to homogeneous hydrogenation and phase-transfer catalysis.

There has been a great deal of attention paid recently to the use of supercritical fluids as substitute media for chemical reactions (*1*), including catalytic reactions (*2,3*). Much of this research has focused on the use of single-phase conditions (particularly for homogeneous catalysis), because it is thereby easier to establish that the reaction is taking place in the supercritical phase. However, for practical purposes it may be preferable to perform the reaction under split-phase (i.e. biphasic) conditions. This paper summarizes several studies of reactions performed in this manner.

Liquid/SCF biphasic systems, where the liquid is essentially insoluble in the SCF, have many advantages over single-phase SCF media, including the ability to separate the catalyst from the product and the ability to use catalysts which have poor solubility in common SCFs. In liquid/SCF biphasic systems, the SCF can be used to deliver the starting material to and/or extract the product from the lower phase. The reaction can be performed while the SCF is present, or if the substrate is soluble in the lower phase then the SCF need not be introduced until the reaction is complete. In either case, it is preferable that the catalyst be soluble only in the lower phase and that the lower phase solvent be insoluble or nearly insoluble in the SCF.

Solid/SCF biphasic systems, which often require a phase-transfer catalyst, allow reactions between a SCF-insoluble reagent and a SCF-soluble reagent. The product can then be isolated, uncontaminated by the SCF-insoluble reagent. In some cases, tuning of the SCF can lead to the extraction of kinetic products, which prevents further conversion to the undesired thermodynamic product.

In solid/gas or liquid/gas biphasic systems, the upper phase is a subcritical inert gas which has insufficient density to dissolve any of the reagents (other than reagent gases). The inert gas is used to modify the properties of the condensed phase in some manner which accelerates the reaction or improves the selectivity. The possibility that subcritical gases can affect reactions in this manner has only recently been discovered.

Liquid/SCF Biphasic Systems

Aqueous/SCF biphasic medium. There are several types of biphasic catalysis currently being used or investigated (Table 1). Each of these systems has inherent advantages and disadvantages. The aqueous and fluorous systems have the disadvantage that the catalyst must be modified to make it hydrophilic or fluorophilic. The fluorous/organic and organic/organic systems have partitioning problems, meaning that the catalyst can have substantial solubility in the upper phase. This leads to significant activity loss after the upper product-containing phase is removed. The organic-containing systems have environmental problems, including the cost of the disposal of the organic solvent, and the cost of separating the organic solvent from the product. Many

Table 1. Biphasic solvent systems being used for catalytic transformations.

Lower phase	Upper phase	Catalyst modification	Leading references
H_2O	Organic	water-soluble	9
H_2O	SCF	water-soluble	4
Fluorous	Organic	fluorous	10,11
Ionic liquid	Organic	none	12-14
Ionic liquid	SCF	none	15
Organic	Organic	none	16

of these problems can be solved or ameliorated by switching to SCFs as the upper phase solvent, usually with water as the lower phase. This approach has now been used for a variety of types of catalysts, including homogeneous (4,5), enzymatic (6,7), and now colloidal catalysts (8).

Rhodium colloids are known to be particularly active catalysts for the hydrogenation of benzene derivatives (17). In collaboration with James (at British Columbia), we explored the possibility of using Rh colloids for arene hydrogenation in aqueous/SCF biphasic media (8). The test substrate, 2-methoxy-4-propylphenol, was selected because it is a model for some of the phenolic groups in lignin which are notoriously difficult to reduce. The hydrogenation proceeded readily in an aqueous/scC$_2$H$_6$ medium (Scheme 1, 88% conversion, substrate:Rh ratio of 50:1, at conditions shown in caption to Figure 1) but did not proceed at all in an aqueous/scCO$_2$ medium, probably because of the pH-drop in the liquid phase (Figure 2). The hydrogenation of other arenes in aqueous/scC$_2$H$_6$ medium was also achieved, even with arenes with negligible solubility in water (conversion shown in Figure 1). The hydrogenation of 2-methoxy-4-propylphenol gave primarily cis-2-methoxy-cis-4-propylcyclo-hexanol, with only 10% of the cis-2-methoxy-trans-4-propylcyclohexanol isomer. The hydrogenation of 4-phenyl-2-butanone had 98% selectivity for hydrogenation at the arene ring only.

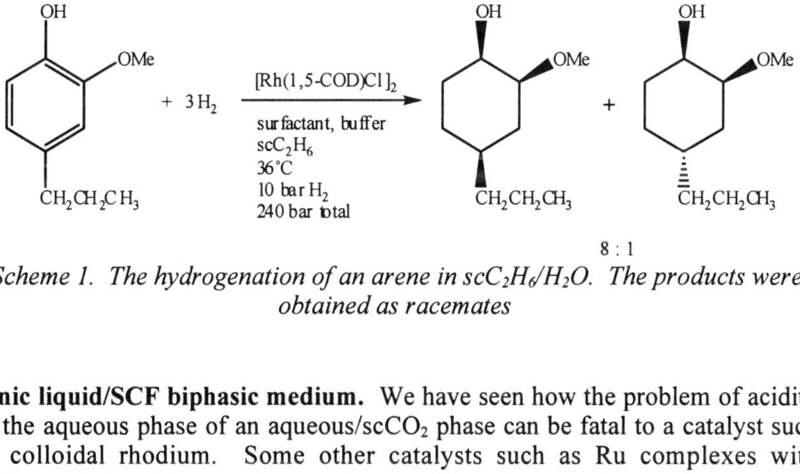

8 : 1

Scheme 1. The hydrogenation of an arene in scC$_2$H$_6$/H$_2$O. The products were obtained as racemates

Ionic liquid/SCF biphasic medium. We have seen how the problem of acidity in the aqueous phase of an aqueous/scCO$_2$ phase can be fatal to a catalyst such as colloidal rhodium. Some other catalysts such as Ru complexes with sulfonated phosphines perform well under such acidic conditions (4). However, for those catalysts which are pH sensitive, one would desire an alternative. One possibility is the use of an aqueous/scC$_2$H$_6$ medium, which solved the problem for the Rh colloidal catalyst. However, both from an environmental and from a safety standpoint, it would be preferable to use scCO$_2$ as the upper phase. Therefore a new lower phase is necessary; a lower phase which, like water, is immiscible with scCO$_2$. Ionic liquids (ILs) such as [bmim]PF$_6$ (structure **1**) have been found to function in this role.

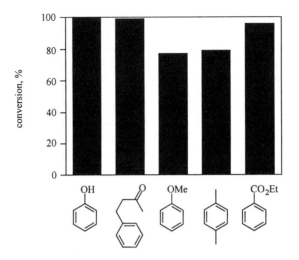

Figure 1. The extent of hydrogenation of arenes in an aqueous/scC₂H₆ medium. Conditions: 36°C, 62 h, 31 mL vessel, 240 bar (10 bar H₂, balance ethane), 9.5 μmol [RhCl(1,5-COD)]₂, 76 μmol [NBu₄]HSO₄, 1.7 mL buffer(0.1 M Na₃PO₄ buffer) and 1.9 mmol arene. Data from reference 8.

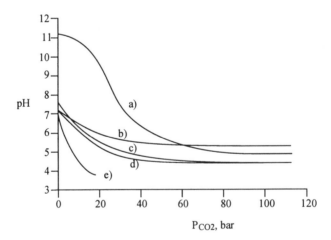

Figure 2. Dependence of the pH of aqueous buffer solutions on the pressure of CO₂ at 36°C in the presence of 9 bar H₂;(8) a) 0.5 M Na₃PO₄ buffer (0.05 M citric acid, 0.2 M boric acid), b) 1.0 M MOPS buffer (3-[N-morpholino]propane-sulfonic acid hemisodium salt), c) 0.1 M Na₃PO₄ buffer (0.05 M citric, 0.2 M boric), d) 0.1 M MOPS, and e) unbuffered water (pH 2.9 at high CO₂ pressure (18)). Reproduced by permission of The Royal Society of Chemistry.

$$\text{Me}\overset{\frown}{-}\text{N}\underset{\smile}{\overset{}{\bigcirc}}\text{N}\text{-Bu} \quad \rceil \ PF_6$$

1

The work of the groups of Brennecke and Beckman (*19*) showed that although CO_2 is soluble in [bmim]PF_6 (0.6 mole fraction CO_2 in the IL phase at 80 bar), the IL is not soluble in the scCO_2 (less than 10^{-5} mole fraction IL in the scCO_2 phase at 138 bar and 40°C). This observation makes it possible for IL/scCO_2 biphasic reactions with extraction of the product by scCO_2 and with no contamination of the product by IL.

Several types of catalysts have been used for hydrogenation reactions in IL (in the absence of CO_2), including several molecular catalysts (RhCl(PPh$_3$)$_3$ (*13*), RuCl$_2$(PPh$_3$)$_3$ (*20*), [bmim]$_3$[Co(CN)$_5$] (*20*), and [Rh(nbd)(PPh$_3$)]PF$_6$ (*12*)) for alkene hydrogenation, a cluster catalyst, [H$_4$Ru$_4$(η^6-C$_6$H$_6$)$_4$][BF$_4$]$_2$, for arene hydrogenation (*21*), and a chiral catalyst, [RuCl$_2$(BINAP)]$_2$•NEt$_3$, for asymmetric hydrogenation (*22*). The last report piqued our interest, especially because the cost of the catalyst makes recycling more appealing, and because the authors reported that the enantioselectivity was pressure independent, a most surprising finding. We chose to test this kind of reaction in combination with scCO_2 extraction of the product (*15*)(*45*).

The chiral catalyst Ru(O$_2$CMe)$_2$((*R*)-tolBINAP) is soluble and active in [bmim]PF_6. While Monteiro et al. (*22*), the authors of the asymmetric hydrogenation study, used atropic acid as their substrate, we chose tiglic acid (Scheme 2), because the enantioselectivity of tiglic acid hydrogenation is usually optimum at low hydrogen concentrations and diffusion rates (atropic acid requires the opposite) (*23*). We anticipated that ILs, which because of their high viscosity should have low diffusion rates, would allow higher enantioselectivity for substrates with those requirements. Indeed, the enantioselectivity of the asymmetric hydrogenation of tiglic acid was higher (*15*) than that found by Monteiro et al. or ourselves for the hydrogenation of atropic acid or its derivatives (Scheme 2, Fig. 3). Also, the enantioselectivity was found to be hydrogen-pressure dependent for tiglic acid. The water content of the ionic liquid was found to have no effect on the enantioselectivity. Extraction of the product, 2-methylbutanoic acid, with scCO_2 was efficient (90% recovery), and no IL or catalyst was observed in the extracted product. Recycling of the catalyst/IL solution by addition of fresh tiglic acid and H$_2$ lead to greater, not lower, enantioselectivity (Figure 3). The manner in which the catalyst solution was recycled is illustrated in Scheme 3. Note that there was no need to modify the catalyst with sulfonated or fluorous groups, as there would have been with aqueous or fluorous biphasic solvent systems.

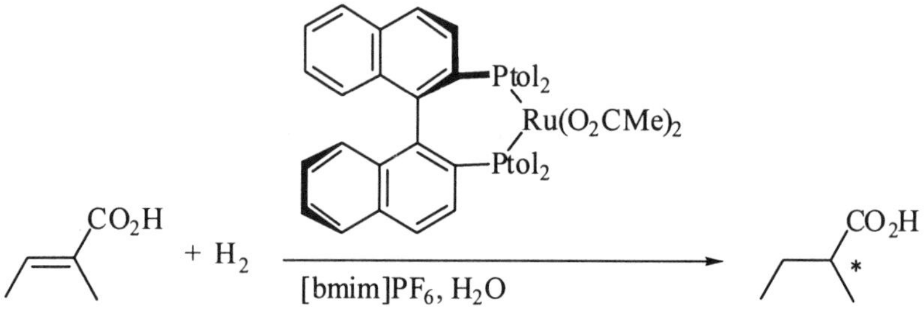

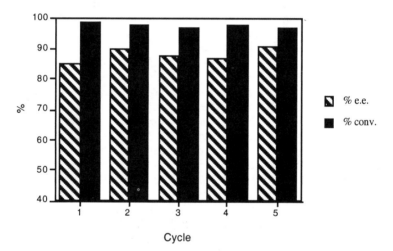

Scheme 2. The hydrogenation of tiglic acid in ionic liquid. The chiral centre in the product is indicated by an asterisk.

Figure 3. The selectivity and activity for tiglic acid hydrogenation of recycled solutions of Ru(O₂CMe)₂(tolBINAP) in [bmim]PF₆ (15). During the fifth cycle, the reaction solution was not stirred..

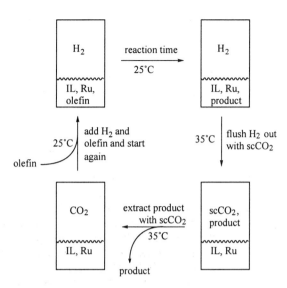

Scheme 3. An illustration of the method for the recycling of ionic liquid/catalyst solutions.

Solid/SCF Biphasic Systems

Solid/SCF reactions mediated by phase-transfer catalysts. Phase-transfer catalysts (PTCs) have been employed as a means of facilitating reaction between solid salts and organic reactants that are present in a supercritical fluid phase. We have investigated several types of reactions in these systems, including halogen exchange, cyanide displacement, and base-promoted alkylation.

Phase-transfer catalysts are molecules with characteristics that give rise to an affinity for both ionic and organic environments. The most common are quaternary ammonium salts, which have organic character derived from the alkyl chains attached to the quaternized nitrogen, but also exhibit behavior characteristic of salts, such as ion exchange. This makes possible the phase-transfer-catalytic cycle, depicted in Figure 4. The diagram shows an inorganic salt, M^+Y^-, that is in contact with a SCF phase containing the organic reactant, R-X. The PTC, Q^+X^-, first exchanges anions with the salt, making a complex that is then able to react with the organic reactant to yield the product and regenerate the catalyst (*24*).

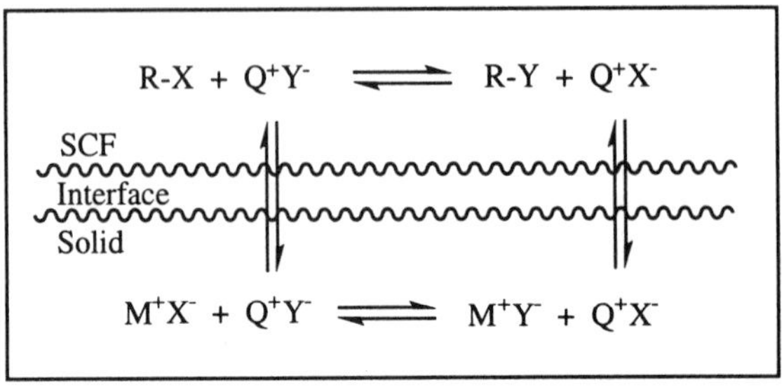

Figure 4. *The phase-transfer-catalytic cycle*.

The first example of PTC incorporated in a SCF system was the reversible halogen exchange between benzyl chloride and potassium bromide to yield benzyl bromide (*25*) (Scheme 4a). The reaction was performed in $scCO_2$ at 50-75°C and 130-200 bar, with tetrahexylammonium chloride as the catalyst.

Another similar reaction that has been studied is the cyanide displacement on benzyl chloride, producing phenylacetonitrile (*26*) (Scheme 4b). This reaction was also carried out in CO_2 at 138 bar, with temperatures at 60°C and 75°C. Tetraheptylammonium chloride was the catalyst, and acetone was used as a cosolvent.

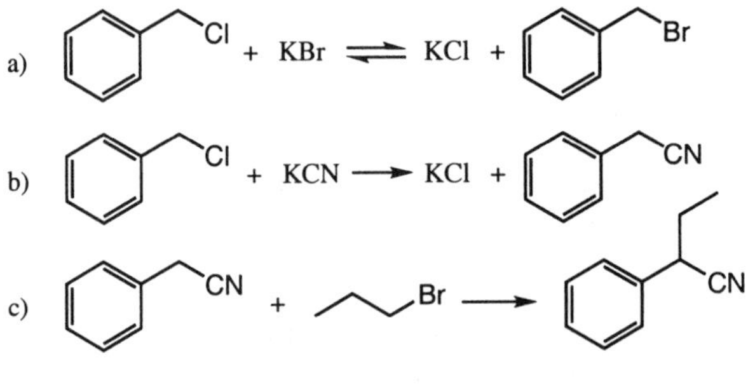

Scheme 4

A third type of phase-transfer-catalyzed reaction that has been investigated in a SCF system is the base-promoted ethylation of phenylacetonitrile in scC_2H_6 (*27*) (Scheme 4c). This reaction could not be performed in CO_2 due to a complicating side reaction. It was also performed at both 60°C and 75°C and at 138 bar, with potassium carbonate as the base and tetrabutylammonium bromide as the phase-transfer catalyst.

The behavior of these reactive systems indicates that the intrinsic reaction is actually taking place on the surface of the solid salt particles rather than in the continuous fluid phase. This assumption is made based on the behavior observed in the reaction of benzyl chloride to phenylacetonitrile.

Although the amount of catalyst was well above the measured solubility limit in scCO$_2$, the rate of reaction continued to increase linearly with the amount of catalyst added to the system. It is therefore believed that the reaction is taking place in a catalyst-rich layer on the surface of the salt particles, termed an "omega-phase."

The activity of the catalysts at concentration levels above their solubility limits facilitates separation of the products from the catalyst. Sufficient catalyst may be used to make reaction rates fast, and then the organic product may be removed from the solids with the SCF, leaving behind the PTC, which may then be recycled. The insolubility of traditional PTCs in scCO$_2$ allows the formation of products that are free from catalyst contamination, which is of paramount interest in the pharmaceutical industry.

Biphasic reaction with simultaneous separation. The dissolving power of SCFs can be easily tuned with density (or cosolvents) to selectively solubilize reaction products without dissolving reactants and catalysts. In addition to the easy solvent removal by depressurization, SCFs provide a substantial advantage over traditional liquid solvents that may indiscriminately solubilize the reactant and catalyst along with the product requiring subsequent separations.

This tunable dissolving power of SCFs provides opportunities to couple a biphasic reaction and a separation in a single process unit. For example, SCFs can be used to remove soluble intermediate products as they are formed before subsequent reaction to unwanted byproducts can take place.

This type of reactive separation has been used in a new synthesis of a precursor to poly(ethylene) terephthalate (PET). Using the precursor, mono 2-hydroxyethyl terephthalate (MHET), offers superior chemical processing as compared to conventional PET production from ethylene glycol and terephthalic acid by reducing heat- and mass-flux loads in the polymerizer, increasing the rate of the polymerization reaction, and reducing the required water removal by half.

MHET was synthesized by the esterification of terephthalic acid (TA) and ethylene oxide (EO) in the presence of a quaternary ammonium salt catalyst. The desired MHET was removed from the involatile bed of terephthalic acid by continuous extraction with supercritical fluid before subsequent reaction to the diester could take place (*28*) (Scheme 5). The SCF, dimethyl ether (Tc = 126.9°C, Pc = 52.4 bar), was tuned with temperature and pressure to readily solubilize and remove the MHET from the bed without also solubilizing the quaternary ammonium salt catalyst or terephthalic acid. A solid mixture (0.2 g) of 95 mol % terephthalic acid and 5 mol % catalyst was loaded into a 6 mm diameter reaction thimble for each experiment. The reaction was run at 130 °C and 70 bar with 0.5 mol % ethylene oxide with respect to dimethyl ether. At a sufficient flowrate of dimethyl ether (5 ml/min pumped at 25 °C and 70 bar), 100 % selectivity to MHET was obtained.

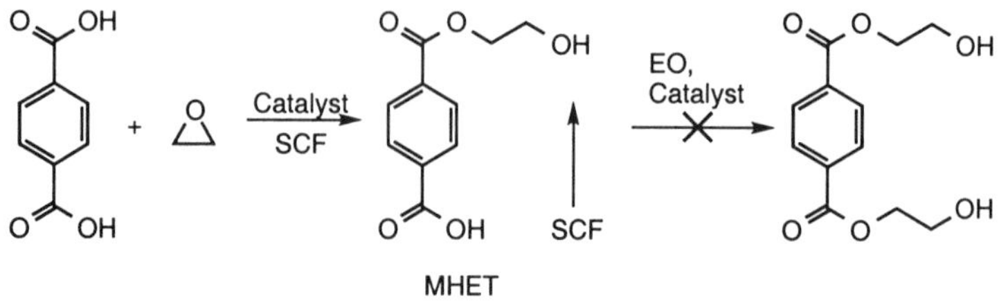

Scheme 5. Selective ethoxylation of terephthalic acid

Solid/Gas or Liquid/Gas Biphasic Systems

Swelling of liquids. In the last two parts of this article, the effect of subcritical inert gases on reactions will be described. When a SCF is used at pressures below its critical pressure, it is no longer technically a SCF but rather a subcritical gas. It has densities too low for it to dissolve any catalysts or reagents other than reagent gases. Organic reagents in the same vessel as a subcritical gas will exist as a condensed phase (liquid or solid) at the bottom of the vessel. If the subcritical gas is chemically inert, then it can only influence the reaction in the condensed phase by dissolving into the condensed phase and modifying the physical properties of the condensed phase.

There is scattered evidence in the literature that subcritical gases can modify the physical properties of condensed phases. The most obvious change is in the volume of a liquid phase when CO_2 pressure is applied. 1,4-dioxane expands more than 12 fold in volume when 71 bar CO_2 is present (*29*). This swelling or expansion of the liquid phase by a subcritical gas is accompanied by other changes:

• The solubility of hydrogen in liquid pentane (15 bar H_2 partial pressure, 50°C) is only 0.012 mol fraction. However, addition of 55 bar of CO_2 causes the H_2 mole fraction solubility of H_2 in the liquid phase to jump 40% (*30*).

• The dielectric constant of pure methanol is 32.6,(*31*) while that for a liquid phase mixture of methanol and CO_2 (78.6 mol% methanol, 35°C, 41.2 bar pressure) is only 25.6 (*32*).

• The normal melting point of p-dichlorobenzene is 53°C, but under 50 bar of C_2H_4 gas, the melting point of the solid drops to 30°C (*33*).

• For the system CO_2/acetone, mutual diffusion coefficients in the liquid phase are predicted to double as the pressure of CO_2 is raised from 10 bar to 70 bar (*34*).

Whether or not an inert subcritical gas can have an effect on the rate of a reaction in a liquid phase was the subject of a study on homogeneous hydrogenation of CO_2 (*35*). The rate of hydrogenation of CO_2 in a liquid NEt_3/MeOH mixture was studied as a function of gas pressure. While increasing the CO_2 pressure increased the rate, this was likely due to the fact that CO_2 was a reagent and may not have been due to the liquid expansion caused by the CO_2 pressure. In order to test the effect of the liquid expansion, inert gases were added (Table 2). When subcritical ethane was added, the rate dropped dramatically. This is believed to be a result of a drop in the dielectric constant of the liquid phase, because a rate drop was also observed if liquid hexane was added. Adding subcritical fluoroform gas did not cause a rate drop, probably because it has a significantly larger dipole moment. In fact, a small but statistically significant rate increase was observed with fluoroform. These results demonstrate that the presence of an inert gas in the upper phase can have a marked effect on the rate of reactions taking place in the lower, condensed, phase.

$$CO_2 + H_2 + {}^1\!/_2 NEt_3 \xrightarrow{\text{RuCl(O}_2\text{CMe)(PMe}_3)_4} {}^1\!/_2 [HCO_2H]_2 NEt_3$$

Table 2. Effect of inert gas on the rate of CO_2 hydrogenation in liquid MeOH.[a]

Added gas (bar)	Yield[b]	TON[c]	TOF[c], h^{-1}
none	0.32	390	770
ethane (40)	0.072	79	160
fluoroform (40)	0.40	455	910
fluoroform (50)	0.40	450	900

SOURCE: Reproduced with permission from reference 35.

[a] Conditions: 50 °C, 0.5 h reaction time, 31 mL vessel, 40 bar H_2, 10 bar CO_2, 2.5 mmol MeOH, 3.6 mmol NEt_3, 3.0 μmol $RuCl(O_2CMe)(PMe_3)_4$.

[b] Moles of HCO_2H per mole of NEt_3; maximum theoretical yield is 2.0.

[c] TON = turnover number = mol HCO_2H per mol of Ru complex. TOF = turnover frequency = mol HCO_2H per mol of Ru complex per hour.

Induced Melting of Solids. Solventless reactions are, environmentally speaking, preferable to reactions with solvents. However, reactions of solids in the absence of a solvent are notoriously slow (there are exceptions (*36*)). We have found that the ability of subcritical gases to lower melting points of organic solids can accelerate some reactions. For example the hydrogenation of solid vinylnaphthalene (mp=62-65°C) (*37*) catalyzed by solid $RhCl(PPh_3)_3$ (Scheme 6) at 33°C is greatly accelerated by the addition of subcritical gaseous CO_2 (*38*).

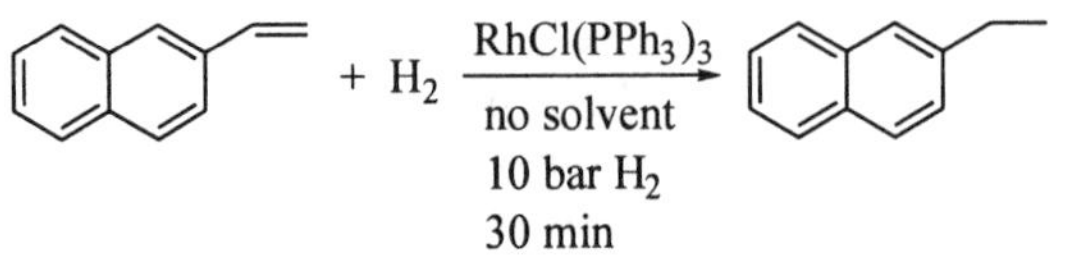

		% conv.	
T,°C		no CO$_2$	56 bar CO$_2$
33		0	52
36		5	73

Scheme 6. The effect of CO$_2$ pressure on the hydrogenation of 2-vinylnaphthalene (38).

Similarly, reactions that form solid products can be difficult to perform without a solvent because the reaction mixture freezes as the reaction approaches completion. For example, the Pt/C-catalyzed hydrogenation of liquid oleic acid (mp = 13-16°C) (39) to form solid stearic acid (mp = 69-70°C) (40) at 35°C proceeds readily until 90% conversion is reached. Thereafter no further conversion is obtained, even after 25 h. In the presence of 60 bar CO$_2$, however, the reaction reaches 97% conversion after only 1 h.

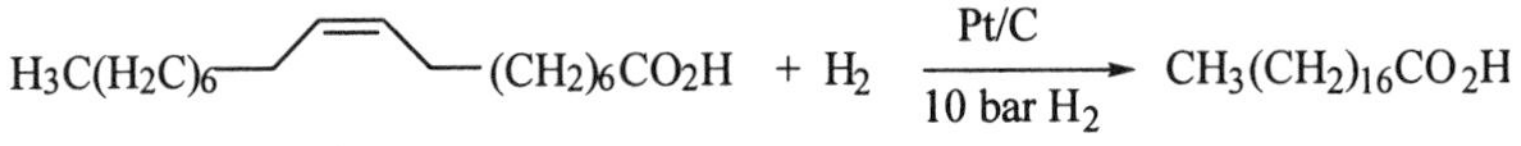

The reason for the rate or yield enhancement is believed to be the melting-point-lowering effect of the subcritical inert gas. This can be explained with reference to the binary phase diagram for a binary mixture of a subcritical gas (component 1) and an organic solid (component 2) having a triple point higher than the critical point of component 1 (Figure 5a). The critical point of the inert gas is shown as C$_1$ and the solid line extending to the left of that point is the boiling curve for component 1. The phase diagram of the organic solid is shown in solid lines on the right side (notice the triple point Tr$_2$ and critical point C$_2$). The lowest temperature at which component 2 could normally melt is its triple point. In the presence of CO$_2$ or another inert gas, the melting point of the condensed phase (the organic compound with CO$_2$ dissolved therein) moves along the dotted SLV line towards the UCEP (upper critical end point). Thus the lowest temperature at which the condensed phase could melt is the temperature of the UCEP. Some solid/inert gas binary mixtures such as p-dichlorobenzene/ethylene and menthol/ethylene have no such UCEP (33) and thus the melting point lowering can be even greater. The composition of the liquid and vapor phases can be determined from an isothermal slice of the phase diagram, as shown in figure 5b.

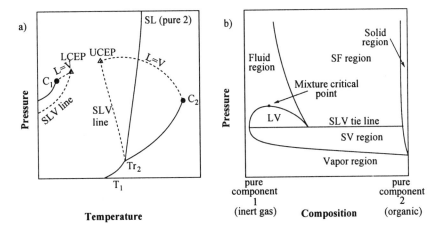

Figure 5. a) The P/T projection of a binary phase diagram (41-44) for a binary mixture of a gas such as CO_2 or C_2H_4 and an organic solid with a triple point higher than the critical point of the gas. b) An isothermal slice of the binary phase diagram at temperature T_1, illustrating the presence of an LV region (a region in which liquid and vapor phases coexist).

Conclusions

Research in the field of reactions in supercritical fluids has been changing from an emphasis on single-phase conditions to a broader approach that includes biphasic conditions. The use of aqueous/SCF or ionic liquid/SCF biphasic catalysis offers an environmentally more benign alternative to the more established methods of aqueous/organic or fluorous/organic biphasic catalysis. Phase-transfer catalysis in solid/SCF biphasic systems allows for clean separation of products from excess reagents and for selective extraction of kinetic products from the solid phase. Finally, we have demonstrated that subcritical pressures of inert gases can alter the rates of reactions taking place in condensed phases. The inert gas affects the rate of the reaction by dissolving in the condensed phase and altering the physical properties of that phase, including its dielectric constant or its melting point.

Acknowledgements. We gratefully acknowledge the assistance and advice of Dr. Brian James (UBC) and Ms. Erin McKoon, Mr. Philip Stalcup, Mr. Yong Huang, and Mr. Scott DeHaai (all of UCD). This material is based upon work supported in part by the EPA/NSF Partnership for Environmental Research under NSF Grant No. 9815320 and in part by the Division of Chemical Sciences, Office of Basic Energy Sciences, Office of Science, U. S. Department

110

of Energy (grant number 069628). This support does not constitute an endorsement by DOE or NSF of the views expressed in this article.

References

1. Jessop, P. G.; Leitner, W., Eds. *Chemical Synthesis using Supercritical Fluids*; VCH/Wiley: Weinheim, 1999.
2. Jessop, P. G.; Ikariya, T.; Noyori, R. *Chem. Rev.* **1999**, *99*, 475-493.
3. Baiker, A. *Chem. Rev.* **1999**, *99*, 453-473.
4. Bhanage, B. M.; Ikushima, Y.; Shirai, M.; Arai, M. *Chem. Commun.* **1999**, 1277-1278.
5. Jacobson, G. B.; Lee, C. T.; Johnston, K. P.; Tumas, W. *J. Am. Chem. Soc.* **1999**, *121*, 11902-11903.
6. Mesiano, A. J.; Beckman, E. J.; Russell, A. J. *Chem. Rev.* **1999**, *99*, 623-633.
7. Aaltonen, O. In *Chemical Synthesis using Supercritical Fluids*; Jessop, P. G., Leitner, W., Eds.; Wiley-VCH: Weinheim, 1999; pp 414-445.
8. Bonilla, R. J.; James, B. R.; Jessop, P. G. *Chem. Commun.* **2000**, 941-942.
9. Cornils, B.; Herrmann, W. A., Eds. *Aqueous-Phase Organometallic Catalysis*; Wiley-VCH: Weinheim, 1998.
10. Cornils, B. *Angew. Chem., Int. Ed. Engl.* **1997**, *36*, 2057-2059.
11. de Wolf, E.; van Koten, G.; Deelman, B.-J. *Chem. Soc. Rev.* **1999**, *28*, 37-41.
12. Chauvin, Y.; Mussmann, L.; Olivier, H. *Angew. Chem., Int. Ed. Engl.* **1995**, *34*, 2698-2700.
13. Suarez, P. A. Z.; Dullius, J. E. L.; Einloft, S.; Desouza, R. F.; Dupont, J. *Polyhedron* **1996**, *15*, 1217-1219.
14. Welton, T. *Chemical Reviews* **1999**, *99*, 2071-2083.
15. Brown, R. A.; Pollet, P.; McKoon, E.; Eckert, C. A.; Liotta, C. L.; Jessop, P. G. *J. Am. Chem. Soc.* **2001**, *123*, 1254.
16. da Rosa, R. G.; Martinelli, L.; da Silva, L. H. M.; Loh, W. *Chem. Commun.* **2000**, 33-34.
17. Januszkiewicz, K. R.; Alper, H. *Organometallics* **1983**, *2*, 1055-1057.
18. Toews, K. L.; Shroll, R. M.; Wai, C. M.; Smart, N. G. *Anal. Chem.* **1995**, *67*, 4040-4043.
19. Blanchard, L. A.; Hancu, D.; Beckman, E. J.; Brennecke, J. F. *Nature* **1999**, *399*, 28-29.
20. Suarez, P. A. Z.; Dullius, J. E. L.; Einloft, S.; deSouza, R. F.; Dupont, J. *Inorganica Chimica Acta* **1997**, *255*, 207-209.

21. Dyson, P. J.; Ellis, D. J.; Parker, D. G.; Welton, T. *Chem. Commun.* **1999**, 25-26.
22. Monteiro, A. L.; Zinn, F. K.; DeSouza, R. F.; Dupont, J. *Tetrahedron-Asymmetry* **1997**, *8*, 177-179.
23. Noyori, R. *Asymmetric Catalysis in Organic Synthesis*; John Wiley and Sons: New York, 1994.
24. Starks, C. M.; Liotta, C. L.; Halpern, M. *Phase-Transfer Catalysis. Fundamentals, Applications, and Industrial Perspectives.*; Chapman & Hall: New York, 1994.
25. Dillow, A. K.; Yun, S. L. J.; Suleiman, D.; Boatright, D. L.; Liotta, C. L.; Eckert, C. A. *Ind. Eng. Chem. Res.* **1996**, *35*, 1801-1806.
26. Chandler, K.; Culp, C. W.; Lamb, D. R.; Liotta, C. L.; Eckert, C. A. *Ind. Eng. Chem. Res.* **1998**, *37*, 3252-3259.
27. Culp, C. W.; Lamb, D. R.; Chandler, K.; Liotta, C. L.; Eckert, C. A. *AIChE Annual Meeting*: Miami Beach, 1998.
28. Brown, J. S.; Lesutis, H. P.; Lamb, D. R.; Bush, D.; Chandler, K.; West, B. L.; Liotta, C. L.; Eckert, C. A.; Schiraldi, D.; Hurley, J. S. *Ind. Eng. Chem. Res.* **1999**, *38*, 3622-3627.
29. Kordikowski, A.; Schenk, A. P.; Van Nielen, R. M.; Peters, C. J. *J. Supercrit. Fluids* **1995**, *8*, 205-216.
30. from data reported in: Freitag, N. P.; Robinson, D. B. *Fluid Phase Equilib.* **1986**, *31*, 183-201.
31. Reichardt, C. *Solvents and Solvent Effects in Organic Chemistry*, 2nd ed.; VCH: Weinheim, 1988.
32. Roskar, V.; Dombro, R. A.; Prentice, G. A.; Westgate, C. R.; McHugh, M. A. *Fluid Phase Equilib.* **1992**, *77*, 241-259.
33. Diepen, G. A. M.; Scheffer, F. E. C. *J. Am. Chem. Soc.* **1948**, *70*, 4081-4085.
34. Dariva, C.; Coelho, L. A. F.; Oliveira, J. V. *Fluid Phase Equilib.* **1999**, *160*, 1045-1054.
35. Thomas, C. A.; Bonilla, R. J.; Huang, Y.; Jessop, P. G. *Can. J. Chem.* **2001**, in press.
36. Tanaka, K.; Toda, F. *Chem. Rev.* **2000**, *100*, 1025-1074.
37. Mowry, D. T.; Renoll, M.; Huber, W. F. *J. Am. Chem. Soc.* **1946**, *68*, 1105-1106.
38. Jessop, P. G.; DeHaai, S.; Wynne, D. C. *Chem. Commun.* **2000**, 693-694.
39. Yoshimoto, N.; Nakamura, T.; Suzuki, M.; Sato, K. *J. Phys. Chem.* **1991**, *95*, 3384-3390.
40. Budavari, S.; O'Neil, M. J.; Smith, A.; Heckelman, P. E.; Kinneary, J. F., Eds. *Merck Index*; 12th ed.; Merck & Co., Inc.: Whitehouse Station, NJ, 1996.

41.   Lamb, D. M.; Barbara, T. M.; Jonas, J. *J. Phys. Chem.* **1986**, *90*, 4210-4215.

42.   McHugh, M. A.; Yogan, T. J. *J. Chem. Eng. Data* **1984**, *29*, 112-115.

43.   McHugh, M.; Krukonis, V. *Supercritical Fluid Extraction*; 2nd ed.; Butterworth-Heinemann: Boston, 1994.

44.   Streett, W. B. In *Chemical Engineering at Supercritical Fluid Conditions*; Paulaitis, M. E., Penninger, J. M. L., Gray Jr., R. D., Davidson, P., Eds.; Ann Arbor Science: Ann Arbor, Michigan, 1983; pp 3-30.

45.   There has now been another report of reactions using an IL/scCO$_2$ biphasic solvent system. Liu, F.; Abrams, M. B.; Baker, R. T.; Tumas, W. *Chem. Commun.* **2001**, 433-434.

Chapter 9

Continuous Polymerizations in Supercritical Carbon Dioxide

Paul A. Charpentier[1,2], Joseph M. DeSimone[1], and George W. Roberts[1]

[1]Department of Chemical Engineering, North Carolina State University, Box 7095, Raleigh, NC 27695–7905
[2]Current address: Department of Chemical and Biochemical Engineering, Faculty of Engineering Science, University of Western Ontario, London, Ontario N6A–5B9, Canada

We have developed a system for the continuous polymerization of various monomers in $scCO_2$ and the continuous removal of polymer particles from high-pressure to ambient conditions. Experiments have been performed with the surfactant-free precipitation polymerization of vinylidene fluoride (VF2) and acrylic acid (AA) utilizing diethyl peroxydicarbonate (DEPDC) as the free-radical initiator for VF2, and 2,2'-azobis(isobutyronitrile) (AIBN) as the free-radical initiator for AA. The PVDF and AA polymers were collected as dry, "free-flowing" powders. Tunable bimodal molecular weight distributions (MWDs) of poly(vinylidene fluoride) (PVDF) were achieved by varying the VF2 concentration. The conversion of VF2 in these polymerizations ranged from 7 to 26%, and the rate of polymerization (R_p) reached a maximum of 27 x 10^{-5} mol/L·s at a VF2 feed monomer concentration of 2.5 mol/L at 75 °C. Homogeneous free-radical kinetics provided a good approximation for the rate of polymerization (R_p) of this heterogeneous polymerization. The PVDF material was characterized by gel permeation chromatography (GPC) and melt flow index (MFI) giving M_w's up to 104 kg/mol, and MFI's as low as 3.0 at 230 °C.

Introduction

Increased environmental concerns and regulations over the use of volatile organic compounds (VOCs) since the late 1980s (e.g., the Montreal Protocol in 1987 and the Clean Air Act amendments of 1990) has caused considerable effort to find environmentally benign solvents and processes for industrial implementation (1). Using batch reactors, DeSimone and coworkers have shown that supercritical carbon dioxide (scCO_2) is a viable and promising alternative medium (T_c= 31.8 °C, P_c= 76 bar) for free-radical, cationic and step-growth polymerizations (2). This work has been summarized in recent reviews (3-5). Indeed, DuPont has earmarked $275 million USD to manufacture Teflon™ commercially in CO_2 by 2006 (6) and have constructed a semi-works plant in Fayetteville, NC for producing several million pounds of Teflon™ annually. The reasons for the intense industrial interest are that CO_2 is inert to highly electrophilic radicals (i.e., no chain transfer to solvent), inexpensive ($100-200/ton), of low toxicity, non-flammable, and environmentally and chemically benign.

In comparison to existing technologies such as aqueous suspension and emulsion techniques for making polymers, CO_2 technology has several significant advantages as it will allow for the elimination of: a) expensive polymer drying steps; b) toxic and hazardous organic solvents; and c) expensive wastewater treatment and disposal steps for removing monomers, surfactants and emulsifiers (7). The main energy-consuming step in aqueous emulsion or suspension polymerization is the heat input required to evaporate water, in order to produce polymer in a dry form. The heat of vaporization (ΔH_v) of water is large, approximately 970 BTU/lb. Moreover, the removal of water to very low concentrations is hampered severely by the high viscosity of the polymer, and the low diffusivity of water through the polymer matrix. In order to dry polymers to low water concentrations, additional energy input is required to continuously renew the surface of the polymer. By contrast, CO_2 plasticizes the polymer to a significant degree, so that the diffusivity of CO_2 out of the polymer is higher than the diffusivities of organic solvents or water, and much less energy is required to expose new surface. Moreover, since CO_2 is supercritical at the end of polymerization, polymer recovery involves separating a solid from a fluid using techniques such as sedimentation or filtration. Although mechanically complex, these techniques have a low intrinsic energy requirement. Table I shows the potential energy savings by 2020 for forming some commodity polymers if their current aqueous processes are converted to a scCO_2 process. This table considers the energy savings from removing the drying operation where well over 100 trillion BTU's/annum savings appears to be possible.

Table I. Potential Energy Savings by Switching to scCO$_2$ Process

Polymer Type	Method of Polymerization	Estimated Production in 2020 (billion lbs)	Energy Savings (x 10^{-12} BTU)
ABS	Emulsion	7	5
SBR	Emulsion	24	16
PVC	Suspension	50	90
PAA	Solution	0.8	9

As industrial interest in using scCO$_2$ as a polymerization medium has grown, several disadvantages of batch processes have been recognized, including: 1) large reactors which are costly at the high pressures of scCO$_2$; 2) difficulty in separating polymer from the supercritical solvent; and 3) difficulty in recycling the CO$_2$ and the unreacted monomer. In order to exploit the advantages of scCO$_2$, we have developed a continuous process for the precipitation polymerization of various monomers in scCO$_2$ (8). Most organic monomers (and initiators) are soluble in liquid and scCO$_2$ (9-11), whereas most polymers are insoluble (12). Therefore, continuous precipitation polymerization in scCO$_2$ should be applicable for polymerization of a wide variety of industrially important monomers. A continuous process requires smaller and hence cheaper equipment for large volume polymers. Moreover, removal of polymer from the system and recycling of monomer and supercritical fluid should be facilitated in a continuous system. A continuous system in scCO$_2$ can also incorporate *in situ* steps to purify the resultant polymer by supercritical fluid extraction (SFE) (1), and to melt process the polymer in CO$_2$, which has the potential to lower the polymer's melt viscosity significantly (13). Indeed, melt viscosities have been reduced in excess of 80% (dependent on polymer system) (32) with a similar reduction in energy for processing.

To date, experiments have been performed with the surfactant-free precipitation polymerization of vinylidene fluoride (VF2) and acrylic acid (AA) utilizing diethyl peroxydicarbonate (DEPDC) as the free-radical initiator for VF2, and 2,2'-azobis(isobutyronitrile) (AIBN) as the free-radical initiator for AA. This work focuses on our efforts to understand the kinetics of VF2 polymerization in scCO$_2$, initiated by the organic peroxide, diethyl peroxydicarbonate (DEPDC). We report on experiments that have been performed at stirring rates from 1300-2700 rpm, initiator inlet concentrations between 8-50 (x 10^{-4}) M, monomer inlet concentrations ranging from 0.4-2.8 M, temperatures between 65-85 °C (at constant CO$_2$ densities of 0.74 g/ml), and residence times from 10 to 50 minutes. These studies will allow us to model the polymerization, and find experimental conditions that allow us to maximize the rate of polymerization (R$_p$) for a desired polymer microstructure and morphology. Until now, there has been very little investigation into the kinetics of free-radical polymerizations carried out in CO_2, either in batch or using a CSTR. As well, very little polymerization data is present in the literature on the PVDF system in particular, and fluorinated monomers in general.

Experimental

Materials. Vinylidene fluoride (VF2) monomer (HFC-1132a) was generously donated by Solvay Reseach, Belgium and used without further purification. Carbon dioxide (SFE/SFC grade) was generously donated by Air Products & Chemicals, Inc. and further purified by passage through columns containing molecular sieves (Aldrich) and copper oxide (Aldrich) to remove excess water and oxygen, respectively. All other chemicals were obtained from the Aldrich Chemical Company.

Initiator Synthesis. The DEPDC initiator was synthesized as previously reported, using water as the reaction medium and extracting the initiator into Freon 113 (HPLC Grade) (17,18). All manipulations of the initiator were performed in an ice bath and the final product was stored in a cold chest at -20 °C. The iodine titration technique, ASTM Method E 298-91, was utilized to determine the concentration of active peroxide in the solution.

Polymerization Apparatus. Figure 1 provides a schematic of the experimental continuous polymerization apparatus. Carbon dioxide and monomer are pumped continuously by syringe pumps (Isco) in constant flow mode and mixed by an 8-element static mixer, before entering the reactor. The initiator solution is also pumped continuously by a syringe pump (Isco) in constant flow mode, and enters the reactor as a separate stream. All feed lines have check-valves to prevent back-flow, thermocouples, and rupture disks for safety in case of overpressurization. The CSTR is an 800 mL Autoclave Engineers (AE) autoclave with a magnedrive to provide mixing of ingredients with an AE dispersimax impeller. The reactor is heated by a furnace, has an installed pressure transducer (Druck) and a thermowell containing a thermocouple (Omega Engineering). The effluent stream leaves the CSTR through the bottom, and is directed by a 3-way ball-valve (HIP) to one of two 280 mL filter housings (Headline) containing 1 μm filters where the solid polymer is collected. Unreacted monomer, initiator and CO_2 pass through the filters and flow through a heated control valve (Badger). This control valve functions as a back-pressure regulator, which controls the reactor pressure at the desired set-point. The effluent stream passes through a water bath to remove unreacted peroxide, while the gaseous CO_2 and monomer are safely vented into a fume-hood. Very low levels of polymer were found in the water bath, so essentially all precipitated polymer was collected on the 1 μm filters.

Reactor Control and Accuracy. Control of the reactor temperature (T) and pressure (P) were excellent during a polymerization, varying within very close tolerances (T= ± 0.3 °C and P= ± 1 bar). A gas chromatograph (GC) (SRI

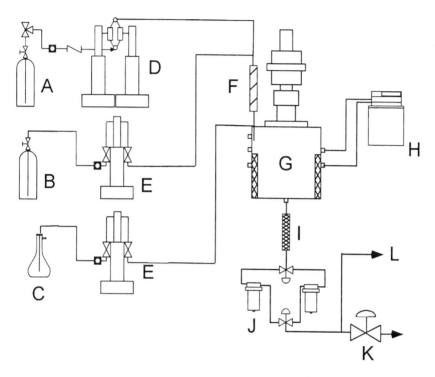

Figure 1.   Schematic of Continuous Polymerization Apparatus. A- CO₂ cylinder; B- Monomer; C- Initiator Solution; D- CO₂ Continuous Pump; E- Syringe Pumps; F- Static Mixer; G- thermostated Autoclave; H- Chiller/Heater Unit; I- Heat Exchanger; J- Steady-State and Non-Steady-State Filters; K-Heated Control Valve; L-GC Analysis.

8610C) was utilized to measure the monomer concentration through sampling the exit stream (after filtration) through a HPLC valve (Valco). The GC column was a silica column while the oven temperature was isothermal at 55 °C with no ramping. Calibration of the GC was performed using CO_2 and VF2 flowrates determined by syringe pumps. Feed rates of initiator and monomer from the syringe pumps were ± 0.1%. Densities of VF2 and CO_2 in the cooled syringe pumps (cooling the syringe pumps by chiller circulators allowed for easier condensing of liquefied gases) and heated reactor were determined from data provided by Solvay for VF2 (Peng-Robinson equation of state) while CO_2 densities were determined from NIST data (19).

Polymer Collection. Collection of "steady-state" polymer normally was started after the reactor had been operating for five mean residence times by switching the reactor effluent from one of the parallel filters shown in Figure 1, to the other. In addition, a novel technique was developed that allowed the polymer powder to be continuously ejected from the collector filter by use of two on/off control valves (Badger). A small 10ml volume between the on/off valves was filled with polymer/CO_2 by actuation of the first valve, which was closed before actuation of the second valve to discharge the polymer rich mixture to a bag filter (McMaster-Carr).

Gel Permeation Chromatography (GPC). All GPC measurements of the PVDF polymer samples were performed by Solvay Research, Belgium on a Waters-Alliance HPLC system with 2x HR5E and 1x HR2E columns using N,N-Dimethylformamide (DMF) modified with LiBr 0.1M at 40 °C. Melt flow indexes were determined with a Kayeness Melt Flow Indexer at 230 °C according to the method of ASTM D-1238.

Results

Attainment of Steady-State. Figure 2 shows the results obtained from a GC analysis of the reactor outlet stream for a typical polymerization run. The effluent VF2 concentration was measured as a function of time onstream, expressed as a multiple of the reactor mean residence time, τ. The fractional conversions of VF2 (monomer conversions) in Figure 2 were calculated from the GC analyses. For a typical polymerization run, steady-state was attained after about 5 τ. Furthermore, if the reactor had been onstream for at least 5 τ, the weight of the polymer collected per unit time was constant, independent of time onstream, confirming the results from the GC analysis.

Experimental Phase Behavior. Under the experimental conditions studied, VF2 and DEPDC were found to be miscible with CO_2, while the formed polymer powder, PVDF, was immiscible in CO_2. These studies were performed in a variable-volume view cell apparatus similar in design to that reported elsewhere (1). Our observations on solubility of PVDF in CO_2 agree with recently

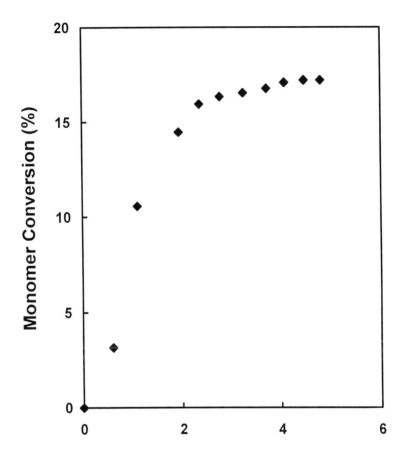

Figure 2.   Attainment of Steady State for Polymerization of VF2. The points
              are experimental data with VF2 analysis obtained by gas
              chromatography. The polymerization conditions were P = 276 bar,
              T = 75 °C, v_{CO2} = 26 g/min, [VF2]$_{INLET}$ = 0.77 M, [EPDC]$_{INLET}$ = 3
              mM, and τ = 21 minutes.

reported data on PVDF/CO$_2$ and PVDF/CH$_2$F$_2$ solubilities (20). This heterogeneous phase behavior defines a precipitation polymerization (21).

Residence Time Distribution (RTD) Studies. To determine whether the experimental reactor functioned as an ideal CSTR, the residence time distribution (RTD) was measured. Both pulse and step techniques were employed (33), with toluene as a tracer. A pulse of toluene was injected into the CO$_2$ stream entering the reactor using a HPLC valve (Valco), and a step input was achieved by starting a constant flow of toluene into the CO$_2$ stream with a syringe pump. The concentration of toluene in the reactor effluent was measured as a function of time by a UV/Vis spectrometer (Shimadzu) at $\lambda = 280$ nm. The dimensionless exit age distribution function, $E(\theta)$, was calculated from the measurements of tracer concentration versus time. The dimensionless time, θ, is the number of reactor volumes of fluid that have flowed through the reactor in time t, i.e., $\theta = t/\tau$, where $\tau = V/\upsilon$, V is the volume of the reactor, and υ is the volumetric flowrate through the reactor. Figure 3 shows a typical normalized RTD trace for the experimental reactor, determined with a pulse injection of tracer. Comparison of the experimental data with $E(\theta)$ for an ideal CSTR clearly illustrates that the behavior of the experimental reactor is indistinguishable from that of an ideal CSTR. For all temperatures between 50 and 90 °C, pressures between 207 and 320 bar, and mean residence times as low as 13 min., the experimental reactor behaved as an ideal CSTR.

Initiator Decomposition Studies. A novel method for the simultaneous determination of the decomposition rate constant, k_d, and the initiator efficiency, f, of a free-radical polymerization initiator in scCO$_2$ was developed (22). The decomposition was carried out in a CSTR in the presence of the radical scavenger galvinoxyl. Combining the mass balances for the initiator and radical scavenger allowed derivation of equation number (1).

$$\frac{2\varepsilon_x l[I]_{IN}}{A_{IN} - A_{OUT}} = \left(\frac{1}{k_D f}\right)\left(\frac{1}{\tau}\right) + \frac{1}{f} \tag{1}$$

The y-intercept of this line is $1/f$ and the slope is $1/(k_D \cdot f)$. Thus, f and k_D can be obtained simultaneously from a set of experiments where the mean residence time, τ, is varied at constant temperature and pressure. The application of Eqn. 1 to the decomposition of ethyl peroxydicarbonate in scCO$_2$ in the presence of galvinoxyl is shown in Figure 4. The experimental data follows the model quite well at each of the four temperatures, and at two different galvinoxyl/initiator feed ratios. The linearity of the experimental data supports several assumptions that were made in the derivation of Equation 1. The most important of these are: a) the decomposition of the initiator is first-order, and; b) the reaction of an initiator radical with the radical scavenger is essentially instantaneous.

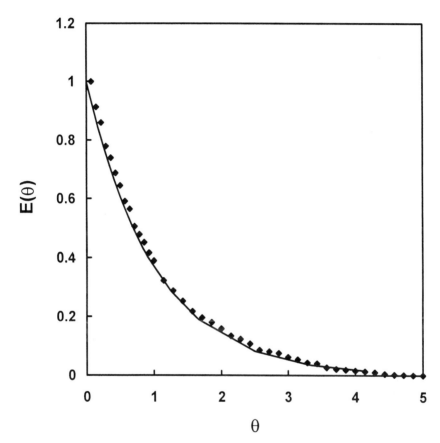

Figure 3.   Comparison of Residence Time Distribution of Experimental
Reactor (♦) to that of an "Ideal" CSTR (—). The experimental
conditions were T= 75 °C, P = 276 bar, ρ_{CO2} = 0.74 g/ml, and τ =
20 mins. Reproduced with permision from Reference 22.
Copyright 2000 Elsevier Science.

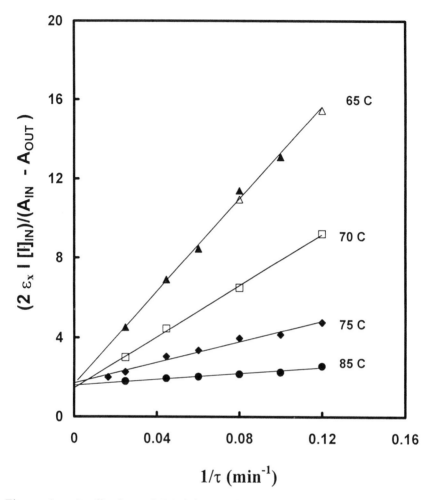

Figure 4.   Application of Model Equation to Decomposition of Diethyl
Peroxydicarbonate in scCO$_2$. The lines are linear least-squares fits
to the points. The experimental conditions were ρ_{CO2} = 0.74 g/ml,
[DEPDC]$_{INLET}$ = 0.3 mM, [galvinoxyl]$_{INLET}$ = 0.6 mM (filled
points) for T = 65 °C(▲), 75°C (♦), 85°C (●); and [DEPDC]$_{INLET}$
= 0.6 mM, [galvinoxyl]$_{INLET}$ = 0.6 mM (unfilled points) for T= 65
°C (Δ),70 °C (□). Reproduced with permision from Reference 22.
Copyright 2000 Elsevier Science.

Table II provides the initiator decomposition rate constants for DEPDC in scCO$_2$ (22). No significant solvent dependence was observed for decomposition of DEPDC in scCO$_2$, compared to literature values using organic solvents at ambient pressures (23). The initiator efficiency found, f = 0.6, is typical for an organic peroxide (21). For the kinetic analysis of PVDF polymerization presented in this paper, the k_d's from Table II were used, while f = 0.6 was used for the initiator efficiency, for all temperatures studied.

Table II. Initiator Decomposition Rate Constants

Temperature ($^\circ C$)	k_D (x 10^4)s^{-1}	f
65	2.4	0.61
70	4.3	0.69
75	10.3	0.59
85	35.1	0.63

SOURCE: Reproduced with permision from Reference 22. Copyright 2000 Elsevier Science.

Effect of Agitation on VF2 Polymerization. The first polymerization experiments dealt with the effect of agitation. Table III provides the effect of stirring rate and agitator type on monomer conversion (X) and on the PVDF weight-average molecular weights (M$_W$). The 1.25" diameter dispersimax™ agitator, a 6-bladed Rushton-type turbine (d/D = 0.42), was studied at agitation rates from 1300 to 2700 rpm. This type of agitator provides mainly radial flow (24). High circulation rates were studied to minimize the formation of deposits, promote good heat transfer (25) and to help suspend the precipitated particles. It is clear from Table 3 that the monomer conversion and M$_W$ were not affected by the agitation rate, for the range investigated. For the lowest stirring rate, 1300 rpm, a different turbine agitator also was studied. This was a four-bladed, 45° pitch, upward-pumping agitator designed to provide a combination of axial and radial flow to help suspend the precipitated polymer particles. The conversion and M$_W$ obtained with this agitator were the same as those with the dispersimax impeller, indicating no effect of agitator geometry on the polymerization, for the conditions studied. For all subsequent experiments reported, the dispersimax impeller was used at a stirring rate of 1800 rpm.

The results shown in Table III lead us to believe that the kinetics were not affected by mixing in this study. Taken in conjunction with the RTD studies that were done with pure CO$_2$ as shown in Figure 3, these results support the assumption that the reactor behaved as an ideal CSTR.

Table III. Effect of Agitation on Conversion and Molecular Weight

Agitation Rate (rpms)	Agitator Type	Monomer Conversion (%)	M_W (x 10^{-3})
1300	D	17.0	24.2
1300	U	17.3	20.3
1800	D	18.0	20.4
2300	D	17.5	-
2700	D	17.7	21.2

D = Dispersimax Impeller™
U = Upward Pumping Impeller

Modeling the Rate of Polymerization
i) Determination of Monomer Order. If the inlet and outlet volumetric flow rates are equal, the mass balance for monomer around an ideal CSTR, operating at steady state, can be written as:

$$R_p = \frac{\left([M]_{IN} - [M]_{OUT}\right)}{\tau} \tag{2}$$

where R_p is the rate of polymerization per unit volume, $[M]_{IN}$ is the concentration of monomer in the inlet stream, $[M]_{OUT}$ is the concentration of monomer in the outlet stream (which is the same as the concentration of monomer in the reactor for an ideal CSTR), and τ is the mean reactor residence time. For this work, $[M]_{OUT}$ was determined gravimetrically by weighing the polymer collected at steady-state, and confirmed by on-line GC analysis of VF2. This allowed the R_p to be determined experimentally, because $[M]_{IN}$ and τ were known.

In "classical" free-radical polymerization, R_p is proportional to the first power of the monomer concentration and to the square root of the initiator concentration (21). Figure 5 provides a plot of R_p versus $[VF2]_{OUT}$ for a series of experiments at 75 °C where the monomer concentration was varied over a wide range and where the inlet and outlet concentrations of DEPDC were constant. This figure illustrates that R_p is nonlinear with the monomer concentration; the slope of the curve decreases as the monomer concentration increases. Table IV summarizes the experimental rate data.

Table IV. $R_p/[M]_{out}$ Values at Different Reactor Temperatures

Temperature (°C)	Pressure (psig)	τ (mins)	$[I]_{IN}$ (mmol/L)	$R_p/[M]_{out}$
65	3400	21.1	2.9	0.58
70	3700	21.2	2.9	0.91
75	4000	21.3	2.9	1.6
80	4300	21.3	2.9	2.3

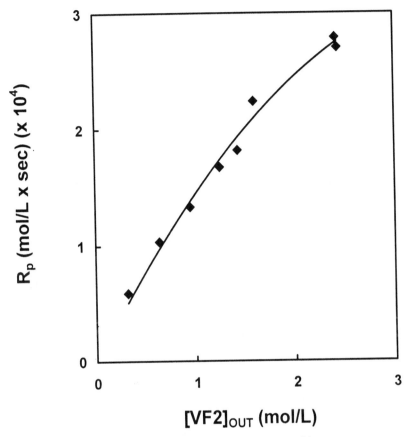

Figure 5. Rate of polymerization versus $[VF2]_{OUT}^{1.0}$ to test first order monomer dependency. The points are experimental data and are fitted with a second order polynomial. The polymerization conditions were P = 276 bar, T = 75 °C, ν_{CO2} = 26 g/min, $[EPDC]_{OUT}$ = 1.3 mM, and τ = 21 minutes.

ii) Determination of Initiator Order. For a CSTR at steady state, with equal inlet and outlet volumetric flow rates, the mass balance for the initiator is:

$$[I]_{IN} \, v_0 - [I]_{OUT} v - r_I V = 0 \qquad (3)$$

where $[I]_{IN}$ is the concentration of the initiator in the inlet stream, $[I]_{OUT}$ is the concentration of the initiator in the outlet stream, and r_I is the rate of consumption of the initiator by decomposition, per unit volume. Assuming a first-order decomposition, $r_I = k_D [I]_{OUT}$, giving:

$$[I]_{OUT} = \frac{[I]_{IN}}{1 + k_D \tau} \qquad (4)$$

Hence, the concentration of the initiator in the reactor, $[I]_{OUT}$, can be calculated from the known inlet concentration, $[I]_{IN}$, the known mean residence time, τ, and the decomposition rate constant, k_D, as provided in Table II.

Figure 6 is a plot of $R_p/[M]_{OUT}$ versus $[I]_{OUT}^{0.5}$, for another series of experiments where the inlet initiator concentration was varied at otherwise constant conditions. Because the initiator concentration changed, the outlet monomer concentration also varied from experiment to experiment. This plot tests both the first-order dependence on the monomer concentration and the square-root dependence on the initiator concentration. The "classical" rate model appears to apply, except for an "offset" on the x-axis, i.e., the line through the data points does not go through the origin. In essence, the data suggest that a threshold initiator concentration of about 10^{-4} mol/L is required to produce a finite rate of polymerization, for the conditions shown.

Figure 6 suggests the presence of an inhibitor, and Figure 5 suggests that the inhibition effect may be associated with the monomer. The model for R_p is based on the reactions shown in Scheme 1.

Scheme 1. Reactions for Modeling the Rate of Polymerization

(1) $\quad I \xrightarrow{\ K_d\ } 2R_i^{\bullet}$ $\qquad$ decomposition of initiator

(2) $\quad P_n^{\bullet} + M \xrightarrow{\ K_p\ } P_{n+1}^{\bullet}$ $\qquad$ propagation

(3) $\quad P_r^{\bullet} + P_s^{\bullet} \xrightarrow{\ K_T\ } P_{r+s}$ $\qquad$ termination (combination and/or disproportionation)

(4) $\quad P_r^{\bullet} + Q \xrightarrow{\ K_Q\ } Pn$ $\qquad$ termination by inhibitor

Rate-Model

The model for R_p is based on the following assumptions (35): (1) the initiation, chain-growth and termination reactions occur in the fluid-phase (i.e.,

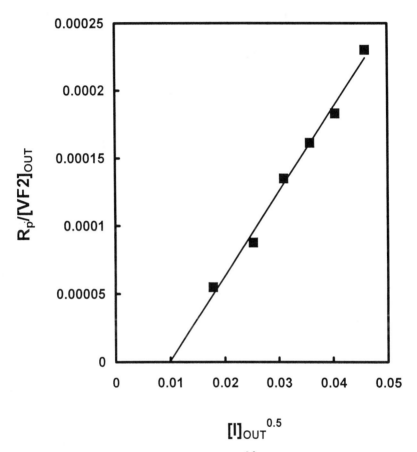

Figure 6.   Plot of $R_p/[VF2]_{OUT}$ versus $[I]_{OUT}^{0.5}$ to test ½ order dependency of initiator concentration. The points are experimental data and the line is a linear least-squares regression fit to the points. The polymerization conditions were the same as Figure 5.

the rates of initiator decomposition, monomer consumption, and "dead" polymer formation in the solid polymer are negligible compared to these rates in the fluid phase), (2) the steady-state approximation (SSA) is valid, (3) rate constants are independent of the chain length, and (4) an inhibitor, Q, is present in the feed. Any free radical that might be formed in reaction 4 above is not effective in initiating new polymer chains. The derived model is provided in eqn. (5).

$$\frac{R_P}{[M]_{OUT}} = \alpha \cdot \left\{ [I]_{out}^{1/2} - \beta + \frac{\beta^2}{2[I]_{out}^{1/2}} \right\} \tag{5}$$

where $\alpha = k_p \left(\dfrac{f \cdot k_D}{k_T} \right)^{1/2}$ and $\beta = (k_Q [Q]/4k_T)/(fk_D/k_T)^{1/2}$

Equation 5 is consistent with the data in Figure 6. At relatively high values of $[I]_{OUT}^{1/2}$, the term $\beta^2/2[I]_{OUT}^{1/2}$ will be small compared to β. In this region, a plot of $R_p/[M]_{OUT}$ versus $[I]_{OUT}^{1/2}$ will be linear, with a slope of α and an intercept of β on the x axis. This is exactly the behavior demonstrated in Figure 6.

In order to determine the robustness of the developed model, experiments were carried out under varying conditions and the data were fitted to Equation 5 by non-linear regression (35). All of the experimental data from this study is provided in Figure 7, a parity plot comparing the experimentally determined R_p's to those predicted from Equation 5. The agreement between the experiments and the rate model is excellent.

First-order dependency for monomer and half-order dependency for initiator is found in "classical" free-radical kinetics. The only reported kinetic data for VF2 polymerization were performed in solution (27). Monomer orders were found approaching 1.8 and the initiator orders reported approximated the classical 0.5 value. Several heterogeneous polymerizations, such as vinyl chloride and acyrlonitrile polymerizations, often show initiator exponents exceeding this classical value (28). This behavior often is attributed to polymer radicals precipitating during the reaction in the nonsolvent environment to form tightly coiled chains which "trap" or "occlude" the radicals. These trapped radicals can still propagate with monomer, but have trouble terminating, hence leading to autoacceleration and initiator exponents greater than 0.5. However, no autoaccelation was observed in these experiments. As well, radical trapping in precipitation polymerizations decreases with increasing reaction temperatures and is normally insignificant above 60 °C (28). Relatively high temperatures were used in this study, i.e. 65-85 °C, so occlusion would be minimized.

Polymer Molecular Weights

Table V provides values of M_W and MFI for some of the PVDF prepared in this research and provides a comparison with commercial material

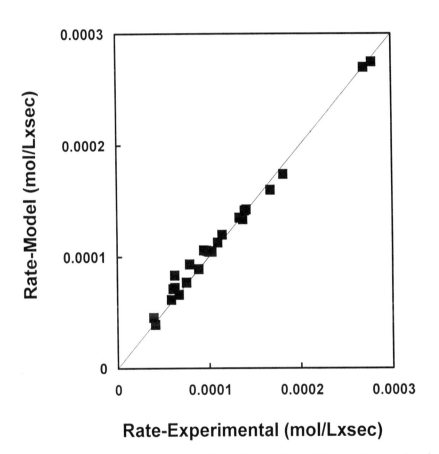

Figure 7.   Parity plot showing the fit of all experimental data to that predicted from Equation 5.

prepared by suspension polymerization. The highest M_W samples prepared in this study are comparable in M_W and MFI to commercial material. A very interesting and important result was found when studying the PVDF molecular weight distributions (MWDs). Figure 8 shows the affect of increasing monomer concentration on the MWDs, which change from unimodal distributions at low monomer concentrations to bimodal distributions at higher monomer concentrations. Hence a novel approach is available to obtain multi-modal material. Although fluoropolymers, such as vinylidene fluoride, provide exhibit excellent physical properties including resistance to weathering, ultraviolet, and chemicals, these fluoropolymers often possess low melt elasticity causing problems in melt fabrication. Additionally, these materials may experience melt fracture surface properties at relatively low extrusion rates and low shear sensitivity can limit their extrudability. Bimodal materials can overcome many of these problems. No other variables such as pressure, temperature, or mean reactor residence time were found to significantly affect the MWD. One possibility for formation of bimodal materials is the partitioning of monomer and initiator into the polymer phase to provide two distinct polymerization environments-i.e. a fluid phase and polymer phase.

Supercritical CO_2 has a unique ability to plasticize polymeric materials, and its excellent transport properties make it an ideal medium for conveying small molecules into even highly crystalline materials (30). Figure 9a provides a scanning electron micrograph (SEM) of the highest M_W PVDF powder obtained during these experiments. Figure 9b shows a higher magnification of a single particle from the same sample. In general, the higher M_W samples contained larger particles, although a distribution of particle sizes was obtained for a given sample. As the SEM figures indicate, the morphology of the powder PVDF particles provides a relatively porous network which should offer little mass-transfer resistance.

Although this kinetic analysis is based on a homogeneous model, the exact nature and mechanism of this precipitation polymerization is still under analysis. Important parameters for kinetic modeling, such as initiator and monomer partition coefficients, are not presently known for this system. A more detailed R_p and MWD model will require this information, as these polymerizations are likely to be two-phase in nature with the extent of polymerization in each phase depending on monomer conversion and other parameters (31).

Table V. Melt Flow Indexes (MFIs) of PVDF Samples

PVDF Type	Mw (GPC)	MFI Temperature (°C)	MFI (grams/10 minutes)
Experimental	15,000-47,000	170	400-0.5
Experimental	104,300	230	2.6
Commercial	>100,000	230	1.4

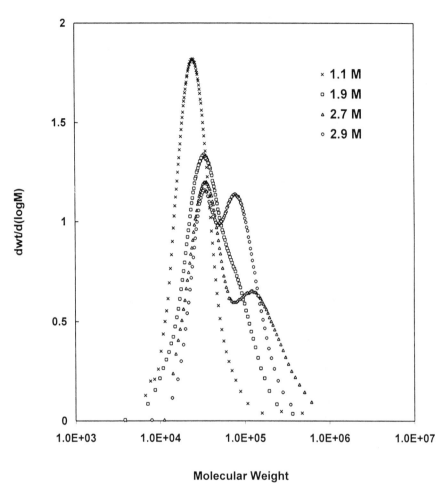

Figure 8.   Gel Permeation Chromatography (GPC) traces of PVDF samples prepared at varying monomer concentrations.   The lines are experimental data.   The experimental conditions are provided in Figure 5.

A)

B)

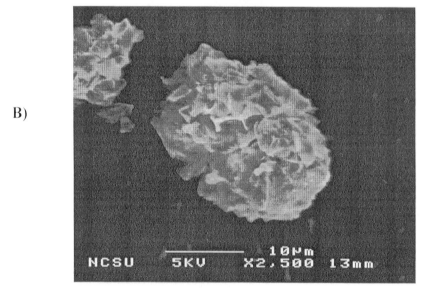

Figure 9.   Scanning Electron Micrographs (SEM) of PVDF powder, a) Low
Magnification; b) High Magnification.

Conclusions

Polymerization of VF2 in $scCO_2$ is heterogeneous (precipitation polymerization). Nevertheless, simple homogeneous chain-growth kinetics provided an adequate description of the rate of this heterogeneous polymerization. The order of the reaction with respect to initiator was 0.5 and the order with respect to monomer was approximately 1.0. Stirring rate and agitator design had no effect on the rate of polymerization, in the range studied. The conversion of VF2 in these polymerizations ranged from 7 to 26 %, and the rate of polymerization (R_p) reached a maximum of 27 x 10^{-5} mol/L·s at a VF2 feed monomer concentration of 2.5 mol/L at 75 °C. The PVDF polymer was collected as a dry "free-flowing" powder, and had M_w's up to 104 kg/mol. "Tunable" bimodal MWDs were achieved.

Notation

ABS	poly (acrylonitrile-butadiene-styrene)
A	absorbance at $\lambda = 762$ nm, dimensionless
A_D	pre-exponential factor, s^{-1}
E_A	activation energy, kJ mol^{-1}
f	initiator efficiency, dimensionless
I	initiator concentration, mol L^{-1}
IN	inlet stream
k_D	initiator decomposition rate constant, s^{-1}
k_P	propogation rate constant, L mol^{-1} s^{-1}
k_T	termination rate constant, L mol^{-1} s^{-1}
k_Q	inhibition rate constant, L mol^{-1} s^{-1}
l	view cell pathlength, m
M	monomer concentration, molar
N_A	Avogadro's constant, mol^{-1}
OUT	outlet stream
P	pressure, bar
PVC	poly(vinyl chloride)
PAA	poly(acrylic acid)
Q	inhibitor, mol L^{-1}
R_p	rate of polymerization, mol L^{-1} s^{-1}
r_I	rate of initiator decomposition, mol L^{-1} s^{-1}
SBR	styrene-butadiene rubber
t	time, s
T	temperature, K
V	reactor volume, m^3

Greek letters

134

ε_x	extinction coefficient, dimensionless
λ	wavelength, nm
θ	dimensionless time
τ	mean residence time, s
ν	volumetric flowrate, $m^3 \, min^{-1}$

Acknowledgement

This work was supported in part by Solvay Advanced Polymers, Inc. through the Kenan Center for the Utilization of Carbon Dioxide in Manufacturing and in part by the STC Program of the National Science Foundation under Agreement No. CHE-9876674. We thank Solvay Research, Belgium for performing GPC analysis of the PVDF samples.

References and Notes

1. McHugh, M. A.; Krukonis, V. J. *Supercritical Fluid Extraction : Principles and Practice*; Second ed.; Butterworth-Heinemann: Boston, 1994.

2. DeSimone, J. M.; Guan, Z.; Elsbernd, C. S. *Science* **1992**, *257*, 945-947.

3. Kendall, J. L.; Canelas, D. A.; Young, J. L.; DeSimone, J. M. *Chem.Rev.* **1999**, *99*, 543-563.

4. Canelas, D. A.; DeSimone, J. M. *Advs. Polym. Sci.* **1997**, *133*, 103-140.

5. Shaffer, K. A.; DeSimone, J. M. *Trends in Polymer Science* **1995**, *3*, 146-153.

6. McCoy, M. In *Chemical & Engineering News*, 1999.

7. Baker, R. T.; Tumas, W. *Science* **1999**, *284*, 1477-1478.

8. Charpentier, P. A.; Kennedy, K.; DeSimone, J. M.; Roberts, G. W. *Macromolecules* **1999**, *32*, 5973-5975.

9. Hyatt, J. A. *J. Org. Chem.* **1984**, *49*, 5097-5101.

10. O'Shea, K. E.; Kirmse, K. M.; Fox, M. A.; Johnston, K. P. *J. Phys. Chem.* **1991**, *95*, 7863.

11. Bartle, K. D.; Clifford, A. A.; Jafar, S. A.; G.F., S. *J. Phys. Chem. Ref. Data* **1991**, *20*, 728.

12. McHugh, M. A.; Krukonis, V. J. In *Encyclopedia of Polymer Science and Engineering, 2nd Ed.*; Mark, H. F., Bikales, N., Overberger, C. G., Menges, G., Kroschwitz , J. I., Eds.; John Wiley & Sons, Inc.: New York, 1989; Vol. 16.

13. Kwag, C.; Manke, C. W.; Gulari, E. *Journal of Polymer Science: Part B. Polymer Physics* **1999**, *37*, 2771-2781.

14. Miller, W. A. In *Chem. Eng.*, 1993.
15. Dohany, J. E.; Humphrey, J. S. In *Encyclopedia of Polymer Science and Engineering*; Mark, H. F., Bilkales, N. M., Overberger, C. G., Menges, G., Eds.; Wiley: New York, 1989; Vol. 17.
16. Russo, S.; Pianca, M.; Moggi, G. In *Polymeric Materials Encyclopedia*; Salamone, J. C., Ed.; CRC: Boca Raton, 1996; Vol. 9.
17. Mageli, O. L.; Sheppard, C. S. *Organic Peroxides*; Wiley-Interscience: New York, 1970.
18. Hiatt, R. *Organic Peroxides*; Wiley-Interscience: New York, 1970.
19. Lemmon, E. W.; McLinden, M. O.; Friend, D. G. In *NIST Chemistry WebBook, NIST Standard Reference Database Number 69*; Mallard, W. G., Linstrom, P. J., Eds.; National Institute of Standards and Technology: Gaithersburg MD, 1998.
20. Lora, M.; Lim, J. S.; McHugh, M. A. *J. Phys. Chem. B.* **1999**, *103*, 2818-2822.
21. Odian, G. *Principles of Polymerization*; 3rd ed.; John Wiley & Sons, Inc.: New York, 1991.
22. Charpentier, P.A., DeSimone, J.M., Roberts, G.W., *Chem. Eng. Science,* **2000**, 55 (22) 5341-5349.
23. Dixon, D. W. In *Polymer Handbook*; Brandrup, J., Immergut, E. H., Grulke, E. A., Eds.; John Wiley & Sons, Inc.: New York, 1999; Vol. II.
24. Geankoplis, C. J. *Transport Processes and Unit Operations*; Third ed.; Prentice Hall: Englewood Cliffs, New Jersey, 1993.
25. Gerrens, H. *CHEMTECH* **1982**, *July*, 434-443.
26. Kamachi, M.; Yamada, B. In *Polymer Handbook-4th Edition*; Brandrup, J., Immergut, E. H., Grulke, E. A., Eds.; John Wiley & Sons, Inc.: New-York, 1999.
27. Russo, S.; Behari, K.; Chengji, S.; Pianca, M.; Barchiesi, E.; Moggi, G. *Polymer* **1993**, *34*, 4777-4781.
28. Bamford, C. H. In *Encyclopedia of Polymer Science and Engineering*; Mark, H., Ed.; John Wiley & Sons: New York, 1988; Vol. 13.
29. Briscoe, B. J.; Lorge, O.; Wajs, A.; Dang, P. *J.Polym. Sci., Part B: Polym.Phys.* **1998**, *36*, 2435-2447.
30. Johnston, K. P.; Condo, P. D.; Paul, D. R. *Macromolecules* **1994**, *27*, 365-371.
31. Xie, T. Y.; Hamielec, A. E.; Wood, P. E.; Woods, D. R. *Polymer* **1991**, *32*, 1098-1111.
32. Royer, J.R.; PhD Dissertation, North Carolina State University**, 2000.**
33. Levenspiel, O; *Chemical Reaction Engineering*, 3rd Ed., pp. 257-269, John Wiley & Sons, New York, NY, 1999.
34. Kim, S. and Johnston, K. P, *Ind. Eng. Chem. Res.*, **1987**, 26, 1206-1213.
35. Charpentier, P.A., DeSimone, J.M., Roberts, G.W, *Ind. Eng. Chem. Res.*, **2000**, 39, (12), 4588-4596.

Chapter 10

An Investigation of Friedel–Crafts Alkylation Reactions in Super- and Subcritical CO$_2$ and under Solventless Reaction Conditions

John E. Chateauneuf and Kan Nie

Department of Chemistry, Western Michigan University, Kalamazoo, MI 49008

This study investigates the Friedel-Crafts alkylation reaction of triphenylmethanol with methoxybenzene (anisole) in supercritical and subcritical carbon dioxide, and under solventless reaction conditions. The reaction was initiated using trifluoroacetic acid to produce triphenylmethlycarbocation as the reaction intermediate. Isolated product yields of the Friedel-Crafts product, p-methoxytetraphenylmethane, are reported. We also present the possibility of using the above reaction as an alternative synthesis in an undergraduate organic laboratory to teach some of the tenets of green chemistry. Additionally, we have investigated the use of benzhydrol, 9-hydroxyxanthene and 9-phenylxanthen-9-ol as potential carbocation sources for supercritical carbon dioxide synthesis and we report those preliminary results.

In recent investigations of electron transfer reactivity of excited-state carbocations in supercritical carbon dioxide (SC CO$_2$), we have demonstrated that stabilized carbocations, e.g., triphenylmethyl cation, xanthenium cation and 9-phenylxanthenium cation, could be generated from their corresponding alcohols in CO_2 under mildly acidic conditions (*1*). *In situ* UV-visible spectroscopy demonstrated that under these conditions the carbocations were stable for days. This stability allowed us to generate photo-excited carbocations and measure their rates of decay using fluorescence spectroscopy. We specifically investigated the pressure, and corresponding solvent density, dependence of the absolute reactivity of electron transfer between excited-state 9-phenylxanthenium cation and toluene, as the electron donor (*2, 3*). During these experiments we noticed that in the presence of toluene the ground state of the carbocation was slowly being consumed. This observation suggested the possibility that an aromatic (Friedel-Crafts) alkylation reaction was occurring under these SC reaction conditions.

Currently, there is significant interest in the development of green chemistry (*5*) and one of the maturing areas of green chemistry is the use of more environmentally friendly reaction media, such as SC fluids (*6*), as alternatives to traditional organic solvents. In this regard, we decided to pursue the practicality of performing Friedel-Crafts (FC) reactions under homogeneous SC reaction conditions. In conventional solvents, FC reactions are typically performed using strong Bronsted acids (H_2SO_4 and HF) or Lewis acids ($AlCl_3$ and BF_3). These reactions often require more than one stoichiometric equivalent of FC catalyst and produce large quantities of waste that are often difficult and costly to separate from the desired product. As a result, traditional FC reactions are far from being atom efficient, nor can they be considered environmentally friendly (*7*). In contrast, the previously mentioned SC CO_2 reaction conditions (*1-3*) seemed promising, in that we were using relatively small quantities of trifluoroacetic acid (TFA) as our protonation source, i.e., FC catalyst, and depressurization methodologies could be employed to recycle the majority of the CO_2 solvent and acid used.

For our initial FC synthesis (*8*), we chose the preparation of *p*-methoxytetraphenylmethane (p-MTPM), see Scheme I, as an example of FC aromatic alkylation. This reaction was chosen for the following reasons: 1) We knew, from previous spectroscopic measurements, that a reaction mixture of triphenylmethanol and TFA (1.0×10^{-3} M and 2.6×10^{-2} M, respectively) were soluble in SC CO_2 at moderate pressures (ca. 100-150 bar) and produced triphenylmethyl cation. 2) Methoxybenzene is extremely reactive toward electrophilic substitution, and 3) p-MTPM is easily identified as an isolated product, m.p. = 201-202°C, and our product yields could be directly compared to a conventional (glacial acetic acid / sulfuric acid solvent) preparation of p-MTPM (*9*).

Scheme 1 presents the mechanism of p-MTPM production. Following initial protonation of triphenylmethanol, the oxonium intermediate undergoes loss of water to produce triphenylmethylcarbocation. In a second step, methoxybenzene undergoes electrophilic substitution by forming an intermediate

Scheme 1

$$(C_6H_5)_3COH \xrightleftharpoons{H^+} (C_6H_5)_3C\overset{+}{O}H_2 \xrightleftharpoons (C_6H_5)_3C^+ + H_2O \quad (1)$$

$$(C_6H_5)_3C^+ + C_6H_5OCH_3 \longrightarrow$$

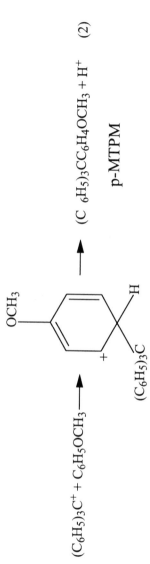

$$\longrightarrow (C_6H_5)_3CC_6H_4OCH_3 + H^+ \quad (2)$$

p-MTPM

sigma complex, followed by p-MTPM production with a proton being released to the medium.

Background information of FC synthesis that use homogeneous SC fluids as the reaction medium is limited. For the specific case of SC CO_2, this is undoubtedly due to very low solubility of conventional FC catalysts in CO_2. FC alkylation reactions have been reported, however, using high-temperature (near-critical) water as a homogeneous reaction medium (10, 11). Eckert and coworkers demonstrated phenol and p-cresol can be alkylated with *tert*-butyl alcohol and 2-propanol in water at 275°C in the absence of added acid catalyst. Therefore, under the near-critical conditions used in those investigations, water acted as both the solvent and catalyst. There is significantly more background regarding the use of SC fluids for heterogeneous catalysis, as described, for example, in the recent review of Baiker (12), and specifically the use of solid acid catalysts for FC reactions, which are described elsewhere (7-8, 13-14).

In this report we present the details of the p-MTPM synthesis and yield of p-MTPM produced under both super and subcritical CO_2 conditions and under solventless reaction conditions. We also report attempts to react three other carbocations of varying structure and stability with anisole and benzene in supercritical CO_2. Additionally, we present the possibility of using the above reaction as an alternative synthesis in an undergraduate organic laboratory to teach some of the tenets of green chemistry.

Experimental

Materials

Triphenylmethanol (97%), methoxybenzene (99%), benzene (99+%), trifluoroacetic acid (99%), benzhydrol (95%), 9-hydroxyxanthene (98%), 9-phenylxanthen-9-ol (98%) and chloroform-d (99.8 atom % D) were purchased from Aldrich and used as received, as was the dichloromethane, Acros Organics (GC grade). Eastman Kodak Co. Chromagram Sheet 13181 (100-micron silica gel) thin layer chromatography plates were used for analytical chromatography and Fisher 28-200 mesh silica gel was used for column chromatography. The GC-MS system used was a Hewlett Packard HP 6890 series GC (HP-5MS column) with a Hewlett Packard 5973 mass selective detector.

Authentic samples of p-MTPM were prepared using a conventional synthesis as described by Wilcox and Wilcox (9). Briefly, in a 250 mL round bottom flask, 3.0 g (0.012 mole) of triphenylmethanol was added to 9.0 mL (0.083 mole) of anisole, followed by 85 mL glacial acetic acid and 8.5 mL concentrated sulfuric acid. The solution was stirred magnetically and refluxed for 1 hour. After cooling, the solution was poured into 100 mL of water and the white precipitate was collected on a Buchner funnel and recrystallized from 1:1 (v:v) toluene:2-propanol which gave 35% yield p-MTPM, m.p. = 198-200°C (9).

Equipment

The apparatus used to deliver CO_2 to either the high pressure reactor or the high pressure optical cell, *vide infra*, contained the following components: a CO_2 gas cylinder, Air Products (SFC grade), fitted with a diptube; and a model 260D Isco high pressure syringe pump, thermostatted with a water jacket and a Fisher Scientific ISOTEMP 1006s heater/chiller re-circulation bath. Connections to the reactor or optical cell were made with HIP 1/16" stainless steel high pressure tubing and high pressure (HIP) line valves fitted with Teflon o-rings.

The high pressure reactor apparatus is described elsewhere (8). Briefly, it was constructed using a custom built cylindrical stainless steel reaction vessel having ca. 79 mL capacity. The vessel was fitted with inlet and exit valves, a Noshok 5000 psi pressure gauge and a HIP rupture disk. When in operation, the reactor was magnetically stirred (stirring bar) and heated using Omega (Omegalux) heating tape that was regulated using a variable voltage regulator. The external temperature of the reaction vessel was monitored using an Omega model HHII digital thermometer and a type K thermocouple.

The high pressure optical cell was constructed of stainless steel, had ca. 3 mL capacity (1.7cm optical path length) and was fitted with Suprasil quartz windows having mechanical seals with lead packing. The design, construction and operational use of our optical cells have been described elsewhere (15). Briefly, a model 901A Heise pressure gauge, Dresser Industries was used to monitor pressure and an Omega Model CN-6070A temperature controller equipped with a cartridge heater and a platinum resistance thermometer was used to control temperature.

Procedure: p-MTPM Synthesis

The standard procedure for the experiments that used CO_2 involved placing known amounts of triphenylmethanol (20.54 mg; 1.0×10^{-3} M), anisole (60.00 mg; 7.0×10^{-3} M) and TFA (0.158 mL; 2.6×10^{-2} M) into the reaction vessel and charging the vessel with pure CO_2 from the Isco pump to a pressure that upon heating would generate a desired reaction pressure. Typically, the reactions were maintained at temperature (± 2°C) for 16-18 hours (see Table I). Following cooling to room temperature the reactor was slowly depressurized into a flask containing CH_2Cl_2. The reaction products from the depressurized vessel were also dissolved in CH_2Cl_2 and the combined volumes reduced for chromatography. It should be noted that the mass of material recovered from the reactor indicated no loss of material during depressurization and transfer. p-MTPM was isolated using silica gel column chromatography with CH_2Cl_2 as the eluent.

TLC, using petroleum ether (30-60) as the developing solvent, was used to analyze fractions. Fractions contained only p-MTPM, as compared to authentic samples ($R_f = 0.3$) were combined and the solvent removed *in vacuo*. Isolated product yields for the experiments performed using super- and subcritical CO_2, the are presented in Table I. The conditions and results of the reaction performed under other than CO_2 conditions are presented in Table II.

Procedure: Other Cations and the Green Chemistry Experiment

The procedure used to generate cations from benzhydrol, 9-hydroxyxanthene and 9-phenylxanthen-9-ol was identical to those described for the p-MTPM synthesis. However, work-up and analysis was accomplished by dissolving the crude reaction mixture in a few milliliters of CH_2Cl_2 and performing GC-MS analysis. The procedure for the green chemistry experiments are described in the green chemistry Results and Discussion Section.

Results and Discussion

Triphenylmethylcarbocation

Table I presents the isolated percent yields of p-MTPM under a variety of reaction conditions. For the reactions performed using CO_2 the fluid density of CO_2 is also given. These fluid densities are of <u>pure</u> CO_2, and were calculated using the method of Angus (*16*). Of course, the phase behavior of <u>pure</u> CO_2 does not accurately describe the reaction mixture. Nevertheless, the calculated CO_2 densities are qualitatively helpful. Spectroscopic or visual observation of the phase behavior of the reaction mixture were performed for two specific reaction conditions, one high density and one low density, *vide infra*.

Our initial experiment was conducted at 115°C. This temperature was chosen to resemble the thermal conditions of the conventional synthesis performed in acetic acid solution, the b.p. of acetic acid is 118°C. The reaction pressure was 229.5 bar, which corresponds to a solvent density (pure CO_2 density) of 0.483 g/cm^3. The critical density of CO_2 is 0.468 g/cm^3 (*17*). Therefore, the reaction mixture density should be, at the very least, near critical.

Table I. Isolated Yields of p-MTPM using Super- and Subcritical CO_2

Temp ($^\circ$C)	Pressure (bar)	Density (g/cm^3)	% Yield	Reaction Time
115	229.5	0.483	3.25	17 hours
70	346.7	0.824	3.62	16 hours
23	208.8	0.929	N/A	17 hours
87	208.8	0.571	3.98	18 hours
91	139.9	0.327	3.62	17 hours
95	84.7	0.154	37.5	16 hours
95	81.3	0.146	36.2	16 hours
95	42.4	0.068	83.2	16 hours

Table II. Isolated Yields of p-MTPM under Solventless Reaction Conditions

Solvent/ Atmosphere	Temp. ($^\circ$C)	Pressure (bar)	Density (g/cm^3)	% Yield	Reaction Time (hrs)
w / N_2	94	1.0	N/A	83.2	17
w / N_2	95	1.0	N/A	86.8	16
Neat/CH_3CO_2H	118	1.0	1.049	3.65	1
Neat/CF_3CO_2H	72.4	1.0	1.480	83.2	1

The isolated percent yield for this reaction was 3.25% (see Table I), which was far below the yield of ~35% obtained in the conventional synthesis. We realized that at this near critical density it was possible that the reactants were not all in a single phase. In an attempt to improve the yield, we increased the fluid density to ca. 0.824 g/cm^3 by lowering the reaction temperature to 70°C at a pressure of 346.7 bar. At a comparable elevated density, both visual inspection and spectroscopic detection (*vide supra*) indicated a single phase for the reaction mixture. Interestingly, at this liquid-like density, the yield of product remained at 3.62%. To determine if elevated temperature was necessary to produce p-MTPM, we performed a reaction at room temperature (23°C) and at 208.8 bar CO_2 pressure. Following our standard work-up procedure, no isolated product was obtained. We estimate this to be less than 1%. This experiment also indicated that there was no significant production of p-MTPM in our reaction vessel prior to charging the vessel with CO_2. To confirm that significant quantities of p-MTMP were not being produced simply upon mixing of the reactants at room temperature, a test tube experiment was performed in which the reactants (20.54 mg triphenylmethanol, 60.00 mg anisole and 0.158 mL TFA) were combined, well mixed and allowed to interact for 1 hour. This experiment also resulted in no isolated p-MTPM. Therefore, albeit in very modest yields, we were confident that the FC alkylation reaction described above was occurring under homogeneous SC CO_2 reaction conditions.

Our next series of experiments were designed to determine if the yield of the FC reaction could be significantly influenced by varying CO_2 fluid density. It is well known (*18*) that at near critical conditions attractive forces between solutes and SC fluids may result in "supercritical solvent clustering". That is, a higher local density of solvent than the bulk density of solvent is found around the solute. Additionally, attractive forces between a solute and a co-solute, or co-reactant may result in an increase in the local composition of the co-solute about the solute. This has been referred to as co-solute "clustering", or enhanced local composition (*18*). These phenomena tend to be most prevalent at near critical fluid densities. In fact, both solvent density augmentation and local composition enhancement have been shown to be greatest at densities below the critical density of mixtures (*18, 19*).

Keeping in mind the possibility of local microstructure influencing the FC reaction, we performed a series of experiments at approximately the same temperature 87-95°C and reaction pressures ranging from 208.8 to 81.3 bar. The corresponding CO_2 density for the reactions ranged from SC density (0.571 g/cm^3) to near-critical, or sub-critical density (0.327 g/cm^3), and to well below the critical density (0.154 and 0.146 g/cm^3). The super and near-critical experiments resulted in 3.98 and 3.62% yields, respectively. Therefore, considering this and the initial data presented above, it would appear that there was no significant influence of local composition on this FC reaction. That is, the yields from the experiment performed at higher, liquid like densities were essentially the same as those obtained from the experiments performed in the near-critical regime. It is in that latter, highly compressible regime where local composition effects are usually more prevalent (*18-20*). At the lower densities (0.154, 0.146 and 0.068 g/cm^3) the percent yield dramatically increased to 37.5, 36.2 and 83.2%, respectively. However, at those reaction conditions the CO_2 density was calculated to be far below the critical density of CO_2. As a result, the reaction would certainly be occurring under subcritical conditions. That was verified visually by observing a liquid phase in a high pressure optical cell. Therefore, the reaction was most likely occurring in essentially neat, CO_2 rich, liquid reactants.

To verify that the dramatic yield increase observed at the lowest CO_2 densities was due to subcritical reaction conditions, experiments were performed using our standard reaction concentrations, however, replacing CO_2 with 1 atmosphere of N_2 at 95°C. The results were approximately 83-87% yield p-MTMP, see Table II. Finally, two experiments (one using TFA and one using acetic acid) were performed, again using same amounts of reactants as in the high-pressure reaction, however, the reactions were performed in a small round-bottom flask, equipped with a condenser, and the reactants were refluxed for 1 hour. In both the TFA and the acetic acid experiments, the reaction mixtures liquefied upon heating. Thus, the reactions were performed in essentially a neat liquid mixture. The results were 83.2% yield in TFA and only 3.65% yield in acetic acid.

The above results demonstrate that it is not always advantageous to perform experiments under homogeneous supercritical reaction conditions. For certain applications, it may be far more beneficial to perform reactions under subcritical, or solventless reaction conditions.

Other Carbocations

In our next series of experiments we attempted to improve the yield of the anisole FC product in supercritical CO_2 by changing the reactivity of the carbocation. The cations generated other than triphenylmethylcation were benzhydryl cation, 9-phenylxanthenium cation and xanthenium cation, see Table III. These cations were generated using trifluoroacetic acid and the respective alcohols; benzhydrol, 9-phenylxanthen-9-ol and 9-hydroxyxanthene. Our stategy and the initial results are presented below. Believing that we could greatly increase the FC product yield by using a carbocation less stable than triphenylmethylcarbocation, we first chose to perform the supercritical CO_2 experiment described above using benzhydrol to generate Ph_2C^+H. However, upon placing the reactants (Ph$_2$CHOH/anisole/trifluoroacetic acid) together in the reactor it was obvious, from a dramatic color change, that product formation occurred upon mixing. GC-MS analysis indicated that the expected FC product (Ph$_2$CHC$_6$H$_4$OCH$_3$) was formed. In an attempt to relieve this "mixing problem" a 10mL beaker was placed in the reactor in order to separate the trifluoroacetic acid from the alcohol until the cell was charged with liquid CO$_2$. Both Ph$_2$CHOH and anisole were place in the beaker and the acid was added to the bottom of the reactor. CO$_2$ was then added, and the reactor brought to temperature. After the reaction time was complete, the cell was depressurized and the beaker was found to be upright and empty. Since the alcohol is a solid, this indicated that the alcohol was dissolved by CO$_2$, and mixing of the alcohol and trifluoroacetic acid occurred in homogeneous CO$_2$. Again, GC-MS analysis showed that the expected FC product was produced. Unfortunately, we can not rule out the possibility that the reaction did not actually occur in the supercritical CO$_2$ phase and that the reaction simply occurred upon depressurization and re-mixing of the reactants in the bottom of the reactor. In addition to the FC product (Ph$_2$CHC$_6$H$_4$OCH$_3$), a second major product was formed in the SC CO$_2$ experiment. The second product generated was determined to be the ether, Ph$_2$CHOCHPh$_2$, formed from the attack of the alcohol precursor (Ph$_2$CHOH) on Ph$_2$C$^+$H followed by loss of a proton. This reaction is presented generically in Scheme 2. The Ph$_2$CHOH/anisole/TFA experiments indicated that Ph$_2$C$^+$H was far too reactive for our purpose. We did however perform one more experiment in hopes of detecting competition in reaction pathways. Benzene was substituted for anisole, since benzene is approximately two orders of magnitude less reactive towards electrophilic substitution than anisole. Before performing a high pressure experiment, a test tube experiment was performed in which the reactants (Ph$_2$CHOH/benzene/TFA) were placed in a test tube at room temperature for two hours. This experiment resulted in production of the FC product (Ph$_3$CH), however no alcohol addition adduct (Ph$_2$CHOCHPh$_2$) was found. Even thought this was not a promising result to demonstrate the <u>need</u> for CO2 as a solvent, we continued and performed an experiment using the beaker separation method in SC CO2 at 92 °C and 300 bar pressure for 17 hours. Intriguingly, in addition to the FC product, the alcohol addition adduct (Ph$_2$CHOCHPh$_2$) was also produced along with higher molecular weight adducts. At this time the additional adducts have not been definitely identified, although they appear to be products formed from free radical addition to the FC product and the initial alcohol adducts.

Next chose to test 9-phenylxanthen-9-ol and, in turn, the generation of 9-phenylxanthenium cation. With this cation we did not observe product upon mixing in either a test tube or in the bottom of the high pressure reactor. Experiments were performed using anisole as the co-reactant under supercritical fluid conditions. Interestingly, GC-MS analysis detected *neither* FC product *nor*

Table III. Carbocations Investigated

Name	Structure
Triphenylmethylcarbocation	Ph_3C^+
Benzhydryl cation	Ph_2C^+H

9-Phenylxanthenium cation
Xanthenium cation

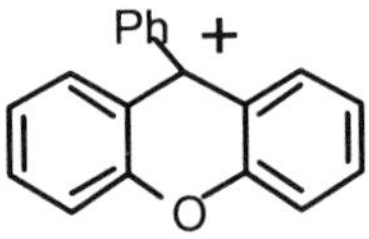

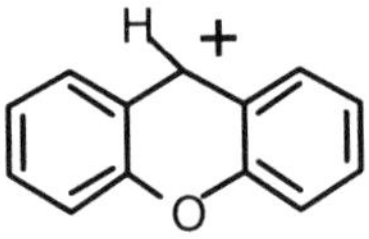

Scheme 2

$$R_3\overset{+}{C} + R_3COH \rightleftharpoons R_3\overset{+}{C}O(H)CR_3 \overset{-H^+}{\rightleftharpoons} R_3COCR_3$$

the ether corresponding to attack of 9-phenylxanthenium cation from 9-phenylxanthen-9-ol. Apparently, 9-phenylxanthenium's stability, or its steric crowding of the cationic center prevented a successful, or irreversible attack at the cationic center. In fact, the residue of the experiment retained the cations characteristic yellow color, which was immediately destroyed by the addition of a small quantity methanol, a fairly good qualitative test of the presence of a stable cation.

Next we chose 9-hydroxyxanthene as the alcohol in order to generate xanthenium cation. This should remove the extra resonance stability and the steric interference of the 9-phenyl substituent of 9-phenylxanthenium cation. In fact, this was the case and again cation and product formation (both FC product and the product formed from attack of the precursor alcohol on xanthenium cation) were observed upon adding the reactants (alcohol/anisole/acid) in a test tube or in the high pressure reactor. Next, a supercritical CO_2 (beaker) experiment was performed and surprisingly, and at this time unexplained, only the corresponding ether product, as produced in Sheme 2, was observed. Experiments were also performed using benzene as the co-reactant. Again the FC product was not observed and only the ether product was detected. Clearly, the reactions in supercritical CO_2, or at least using the supercritical CO_2 procedure, result in different chemistry than that observed under neat room temperature conditions, and these interesting results should be fully examined. In order to do this properly, our reactor should be modified with on line analysis. However, back to the focus of the current work, it is unfortunate that our current attempts of using homogeneous supercritical CO_2 as a practical solvent for FC alkylations have not been successful.

Interestingly, however, when both the FC and the alcohol / cation adduct were formed, an increase in the ratio of FC product to the alcohol ether adduct was detected in the supercritical CO_2 experiments. This indicated that under homogeneous supercritical CO_2 conditions the FC pathway was able to compete with the attack of the carbocation by the alcohol precursor. We are continuing to look for other promising FC systems.

Green Chemistry Experiment

We believe that the experiment presented in the last entry of Table II has excellent potential for the development of a green chemistry undergraduate laboratory experiment. This experiment would demonstrate to students the relatively new chemical concepts of green chemistry, which are now being implemented in nearly all sectors of the chemical industry world wide. The green chemistry concepts that will be demonstrated are designing new synthetic methods that not only result in higher yields of product compared to conventional synthesis, but also reduce quantities of reactant, auxiliary solvents and waste. This experiment will also demonstrate a reduction in the use of corrosive materials, which in turn relates to lower environmental and worker health risks, the ability to recycle materials and in some cases the reduction of energy, time and labor costs.

First, a conventional FC alkylation synthesis of p-MTPM as described in the undergraduate organic laboratory text of Wilcox and Wilcox (9) will be presented. The second edition of the text published in 1995 describes two possible procedures for the p-MTPM experiment. In both cases, experiments are performed using 250 mL reaction vessels, the first an Erlenmeyer flask and the second a round bottom flask. The reactants were identical in quantity and were 3.0 g of Ph$_3$COH, 9 mL of anisole, 85 mL glacial acetic acid and 8.5 mL sulfuric acid. The first experiment was performed by simply mixing the reactants together and allowing the to sit in a hood for 2-7 days. The work-up consisted of pouring the product mixture into a 100 mL of water and collecting the precipitate using a Buchner funnel, with a wash of a few milliliters of diethylether. Recrystallization from 1:1 (v:v) toluene:2-propanol resulted in 45% of the theoretical yield. The second option was to accelerate the reaction by refluxing the reactant for 1 hour, followed by the same work-up procedure which resulted in a 35% yield of p-MTPM.

Our initial experiment presented in Table II, was performed using a 50 mL round bottom and used the following quantities of reactant: 200 mg Ph$_3$COH, 0.60 mL anisole and 1.60 mL TFA. The reactants were refluxed for 1 hour and then the TFA, containing remaining quantities of anisole and a small quantity of water, was distilled from the reaction vessel by simple distillation. The volume of distillate recovered was 1.60 mL and GC-MS analysis determined that 1.30 mL of the distillate was TFA. Remaining in the round bottom flask was crude, but relatively uncontaminated p-MTPM recovered in approximately 70% yield. Without the additional labor of purifying the distillate, i.e., the FC catalyst (1.30 mL mixture) was reused with a fresh quantity of Ph$_3$COH and anisole reactants and the experimental procedure was repeated. The second experiment again gave ~70% yield p-MTPM and ~1.60 mL distillate, however this time the distillate contained 1.13 mL TFA. The recycling experiment was repeated for a total of four runs with a consistent ~70% yield p-MTPM and a total reduction of FC catalyst of a little less than 50%.

The quantities of the materials used and the waste generated (which are primarily the auxiliaries in the conventional synthesis) of each method presented should be pointed out to the students and then compared to the respective yields. Secondly, an atom efficiency calculation of the reactions, including auxiliaries, should be performed which will drive home the point of the unnecessary generation of waste. Thus, in an extremely simple experimental procedure and with relatively minimal concept lecturing, the students will clearly appreciate the developing concepts of green chemistry.

Conclusions

In this work, we first presented isolated product yields of the FC alkylation reaction between triphenylmethanol and anisole performed under supercritical CO_2 and subcritical CO_2 reaction conditions. It was clear from visual observations (high pressure optical cell) that at the highest CO_2 fluid density conditions used in this investigation, the FC reaction occurred under homogeneous conditions. Unfortunately, under these conditions the product yield was low, ~3-4%. Mixed results were obtained using the other alcohols as carbocation precursors. The preparative experiments performed using Ph_3COH under neat liquid conditions, see Table II, demonstrated that FC reactions can be performed very efficiently using neat reaction conditions when TFA is used as the FC catalyst. That observation led us to consider using the neat Ph_3COH/anisole/TFA system as an educational tool to teach undergraduates some of the tenets of green chemistry.

Literature Cited

1. Jin, H. MA Thesis, Western Michigan University, Kalamazoo, MI, 1999.
2. Jin, H.; Chateauneuf, J. E., Reaction of Excited State Carbocations in Supercritical Carbon Dioxide, Annual Meeting of the American Institute of Chemical Engineers, Dallas, Texas, Nov. 2, 1999.
3. Jin, H.; Brennecke, J. F.; Chateauneuf, J. E. Unpublished results.
4. Clark, J. Editorial, *Green Chemistry* **1999**,*1* (1), pp. G1 – G2.
5. Anastas, P. T.; Warner, J. C. *Green Chemistry: Theory and Practice*, Oxford University Press, Oxford, 1998.
6. Noyori, R., ed., 1999, *Chemical Reviews*, Thematic Issue on Supercritical Fluids, American Chemical Society, Washington, DC.
7. Clark, J. H. *Green Chemistry* **1999***1*(1), 1-8.
8. Chateauneuf, J. E.; Nie, K. *Adv. Env. Res.* **2000**, *4*, 307-312.
9. Wilcox, A.; Wilcox, B. *Experimental Organic Chemistry. A Small Scale Approach*; 2nd ED., Prentice-Hall: Englewood Cliffs, NJ, 1995.
10. Savage, P. E. *Chem. Rev.* **1999**,*99* (2), 603-621.
11. Chandler, K.; Deng, F.; Dillow, A. K.; Liotta, C. L.; Eckert, C. A. *Ind. Eng. Chem. Rev.* **1997**,*36* (12), 5175-5179.
12. Baiker, A.*Chem. Rev.* **1999**, *99*, (2), 453-473.

13. Gao, Y.; Liu, H-Z.; Shi, Y-F.; Yuan, W.-K. The 4th International Symposium on Supercritical Fluids, Sendai, Japan, 1997, pp. 531-534.
14. Hitzler, M. G.; Small, F. R.; Ross, S. K.; Poliakoff, M. *Chem. Commun.* **1998,** 359-360.
15. Roberts, C. B.; Zhang, J.; Brennecke, J. F.; Chateauneuf, J. E. *J. Phys. Chem.* **1993,** *97* (21), 5618-5623.
16. *International Thermodynamic Tables of the Fluid State: Carbon Dioxide;* Angus, S.; Armstrong, B.; de Reuck, K. M., Eds.; Pergamon Press, Oxford, 1976.
17. *Lange's Handbook of Chemistry, 13th Ed.;* Dean, J. A., Ed.; McGraw-Hill: NY, 1985; P. 9-182.
18. Brennecke, J. F.; Chateauneuf, J. E. *Chem. Rev.* **1999,** *99*(2), 433-452.
19. Roek, D. P.; Chateauneuf, J. E.; Brennecke, J. F. *Ind. Eng. Chem. Res.* **2000,** *30*, 3090.
20. Zhang, J.; Connery, K. A.; Brennecke, J. F.; Chateauneuf, J. E. *J. Phys. Chem.* **1996,** *100* (30), 12394-12402.

Chapter 11

Lewis Acid Catalysis in Clean Solvents, Water, and Supercritical Carbon Dioxide

Shū Kobayashi and Kei Manabe

Graduate School of Pharmaceutical Sciences, The University of Tokyo, Hongo, Bunkyo-ku, Tokyo 113–0033, Japan

New types of Lewis acids as water-compatible catalysts have been developed. Metal salts such as rare earth metal triflates were found to work as Lewis acids in carbon–carbon bond-forming reactions in the presence of large amounts of water. These salts can be recovered after the reactions and reused. Furthermore, Lewis acid–surfactant-combined catalysts, which can be used for reactions in water without using any organic solvents, have been also developed. Finally, Lewis acid catalysis in supercritical carbon dioxide has been successfully performed. These investigations will contribute to reducing use of harmful organic solvents and to development of environmentally friendly Lewis acid catalysis.

Lewis acid catalysis has been of great interest in organic synthesis.(*1*) While various kinds of Lewis acid-promoted reactions have been developed and many have been applied in industry, these reactions must be generally carried out under strictly anhydrous conditions. The presence of even a small amount of water stops the reaction, because most Lewis acids immediately react with water rather than substrates. In addition, recovery and reuse of the conventional Lewis acids are formidable tasks. These disadvantages have restricted the use of Lewis acids in organic synthesis. From a viewpoint of recent environmental consciousness, however, it is desirable to use water instead of organic solvents as a reaction solvent, since water is a safe, harmless, and environmentally benign solvent.(2) In addition, it is not needed to dry solvents and substrates for the reactions in aqueous media. In the course of our investigations to develop new synthetic methods, we have found that rare earth metal triflates (Sc(OTf)$_3$, Yb(OTf)$_3$, etc.) and some other metal salts can be used as Lewis acids in aqueous media (water-compatible Lewis acids). Furthermore, a new type of a Lewis acid, a Lewis acid–surfactant-combined catalyst, has been developed for the Lewis acid-catalyzed reactions in water without using any organic co-solvents. In this article, our research work on use of the Lewis acid catalysts in aqueous solvents is overviewed.(3) As an extension of this work, we have developed some Lewis acid-catalyzed reactions in supercritical carbon dioxide, which is also regarded as an environmentally friendly solvent.

Water-Compatible Lewis Acids

The titanium tetrachloride (TiCl$_4$)-mediated aldol reaction of silyl enol ethers with aldehydes was first reported in 1973.(4) The reaction (Mukaiyama aldol reaction) is notably distinguished from the conventional aldol reactions carried out under basic conditions; it proceeds in a highly regioselective manner to afford crossed-aldol adducts in high yields.(5) Since this pioneering effort, several efficient activators (Lewis acids) have been developed to realize high yields and selectivities, and now the reaction is considered to be one of the most important carbon–carbon bond-forming reactions in organic synthesis.(6) However, these reactions are usually carried out under strictly anhydrous conditions because the Lewis acids are decomposed rapidly even in the presence of a small amounts of water.

A disadvantage of using anhydrous conditions is that substrates which contain water of crystallization or are available as aqueous solutions cannot be used directly. On the other hand, we found that lanthanide trifluoromethanesulfonate (lanthanide triflates, Ln(OTf)$_3$) functioned as water-compatible Lewis acid catalysts. Lanthanide compounds were expected to act as strong Lewis acids because of their hardness and to have strong affinity toward carbonyl oxygens. Furthermore, their hydrolysis to produce metal hydroxides or oxides and proton was postulated to be thermodynamically disfavored according to their hydrolysis constants.(7) In fact, while most metal triflates are prepared under strictly anhydrous conditions, Ln(OTf)$_3$ are reported to be prepared in aqueous solution.(8) The usefulness of these Lewis acids was first demonstrated in the hydroxymethylation of silyl enol ethers by using commercial aqueous solution of formaldehyde.(9) Among the $Ln(OTf)_3$ tested for the aldol reaction, $Yb(OTf)_3$ was found to be the most effective catalyst (Eq. 1). It is noted that only a catalytic amount of $Yb(OTf)_3$ was required to complete the reaction.

$Ln(OTf)_3$, especially $Yb(OTf)_3$, also activate aldehydes other than formaldehyde in aldol reactions with silyl enol ethers in aqueous solvents such as water–THF.(10) One feature in the present reactions is that water-soluble aldehydes, for instance, acetaldehyde, acrolein, and chloroacetaldehyde can be used directly for the reactions with silyl enol ethers to afford the corresponding crossed-aldol adducts in high yields. Another striking feature of $Ln(OTf)_3$ is that it is very easy to recover them from the reaction mixture. $Ln(OTf)_3$ are more soluble in water than in organic solvents such as dichloromethane. Almost 100% of $Ln(OTf)_3$ was quite easily recovered from the aqueous layer after the reaction was completed and it could be reused. The reactions are usually quenched with water and the products are extracted with an organic solvent. The catalyst is in the aqueous layer and only removal of water gives the catalyst that can be used for the next reaction. Taking into account the facile reuse of the catalysts, $Ln(OTf)_3$ are expected to solve some severe environmental problems induced by conventional Lewis acid-promoted reactions in industrial chemistry.(11)

In the present $Ln(OTf)_3$-catalyzed aldol reactions in aqueous media, the amount of water strongly influences the yields of the aldol adducts. The effect of the amount of water on the yield in the model reaction of benzaldehyde with silyl enol ether **2** in the presence of 10 mol % $Yb(OTf)_3$ in THF was investigated (Eq. 2). The best yields were obtained when the ratios of water in THF were 10–20% (v/v). When the amount of water increased, the yield began to decrease. The reaction system became two phases when the amount of water increased, and the yield decreased. Only 18% of the product was isolated in 100% water. On the other hand, when water was not added or 1–5 eq. of water were added to $Yb(OTf)_3$, the yield of the desired aldol adduct was also low (ca. 10% yield). The yield improved as water was increased to 6–10 eq., and when more than 50 eq. of water were added, the yield improved to more than 80%. These results indicate that water acts as not only a co-solvent but also a kind of an activator in the reaction.

PhCHO + [structure: 1-(trimethylsilyloxy)cyclohexene, OSiMe$_3$] $\xrightarrow[\text{H}_2\text{O–THF, rt}]{\text{Yb(OTf)}_3 (10 \text{ mol }\%)}$ [structure: aldol product, Ph–CH(OH)–cyclohexanone] (Eq. 2)

2

Scandium triflate (Sc(OTf)$_3$) was also found to be an effective catalyst in aldol reactions in aqueous media.(*3b*) For many cases, Sc(OTf)$_3$ was more active than Yb(OTf)$_3$ and Y(OTf)$_3$ as expected from the smaller ionic radius of Sc(III).

In order to search other Lewis acids which can be used in aqueous solvents, group 1–15 metal chlorides, perchlorate, and triflates were screened in the aldol reaction of benzaldehyde with silyl enol ether **1**.(*12*) Not only Sc(III), Y(III), and Ln(III) but also Fe(II), Cu(II), Zn(II), Cd(II), and Pb(II) were found to work as Lewis acids in aqueous media. From these results, we noticed a correlation between the catalytic activity of the metal cations in aqueous media and hydrolysis constants (K_h) and exchange rate constants for substitution of inner-sphere water ligands (water exchange rate constant (WERC)).(*7*) The metal compounds which were active in the aldol reaction have pK_h values from about 4 (4.3 for Sc(III)) to about 10 (10.08 for Cd(II)) and WERC greater than 3.2 x 10^6 M^{-1}s^{-1}. Cations are generally difficult to hydrolyze when their pK_h values are large. In the case that pK_h values are less than 4, cations are easily hydrolyzed to produce certain amounts of proton. Under these conditions, silyl enol ethers decompose rapidly. On the other hand, in the case that pK_h values are more than 10, the Lewis acidity of the cations are too low to catalyze the aldol reaction. WERC values should be large for effective Lewis acids in aqueous media, because large WERC values secure fast exchange between hydrating water molecules and an aldehyde which must coordinate to the metal cation to be activated. "Borderline" elements are Mn(II), Ag(I), and In(III), whose pK_h and WERC values are close to the criteria limits.

Judging from these findings, the mechanism of Lewis acid catalysis (for example, aldol reactions of aldehydes with silyl enol ethers) in aqueous solvents can be assumed to be as follows. When metal compounds are added to water, the metals dissociate and hydration occurs immediately. At this stage, the intramolecular and intermolecular exchange reactions of water molecules frequently occur. If an aldehyde exists in this system, there is a chance for the aldehyde to coordinate to the metal cations instead of the water molecules and the aldehyde is then activated. A silyl enol ether attacks this activated aldehyde to produce the corresponding aldol adduct. According to this mechanism, it is expected that many Lewis acid-catalyzed reactions should be successful in aqueous solution. Although the precise activity as Lewis acids in aqueous media cannot be quantitatively predicted by pK_h and WERC values, these results have shown the possibility of using several promising metal compounds as Lewis acid catalysts in water.

We have also found that these water-compatible Lewis acids such as $Ln(OTf)_3$ can be used in other reactions including allylation of carbonyl compounds with tetraallyltin (13) and Mannich-type reactions of aldehydes, amines, and alkenyl methyl ethers in water-containing solvents.(14)

Catalytic Asymmetric Aldol Reactions in Aqueous Media

Catalytic asymmetric aldol reactions provide one of the most powerful carbon-carbon bond-forming processes affording synthetically useful, optically active β-hydroxy carbonyl compounds.(15) Chiral Lewis acid-catalyzed reactions of aldehydes with silyl enol ethers are the most convenient and promising methods, and several successful examples have been reported since the first chiral tin(II)-catalyzed reactions appeared in 1990.(16) Common characteristics of these catalytic asymmetric reactions include (i) the use of aprotic anhydrous solvents such as dichloromethane, toluene, and propionitrile, and (ii) low reaction temperatures (−78 °C), which are also employed in many other catalytic asymmetric reactions.

In the course of our investigations to develop new chiral catalysts and catalytic asymmetric reactions in aqueous media, we focused on several elements whose salts behave as Lewis acids in such media. After screening several chiral Lewis acids which could be used in aqueous solvents, a combination of $Cu(OTf)_2$ and a bis(oxazoline) ligand was found to give good enantioselectivity (Eq. 3).(17) It should be noted that the reaction of benzaldehyde with (Z)-3-trimethylsiloxy-2-pentene in dry ethanol or dichloromethane in the presence of the chiral catalyst resulted in a much lower yield and selectivity.

We have also developed another example of catalytic asymmetric aldol reactions in water-containing solvents using a combination of a chiral crown ether and $Pb(OTf)_2$ (Eq. 4).(18) In this case, size fitting between Pb(II) cation and the crown ring is an important factor for high enantioselectivity.

Although $Ln(OTf)_3$ are the first Lewis acids which were found to catalyze aldol reactions in aqueous media, it had been difficult to realize asymmetric versions of $Ln(OTf)_3$-catalyzed reactions in such media. Quite recently, we have reported the first example of this type of reactions using a D_2-symmetric chiral bis-pyridino-18-crown-6 (Eq. 5).(19) In the reaction of benzaldehyde with 1 in water–ethanol (1/9), $Ln(OTf)_3$ having large metals such as La, Ce, Pr, and Nd gave the aldol adduct with high diastereo- and enantioselectivities, showing that size-fitting between the crown ether and the metal cations is again an important

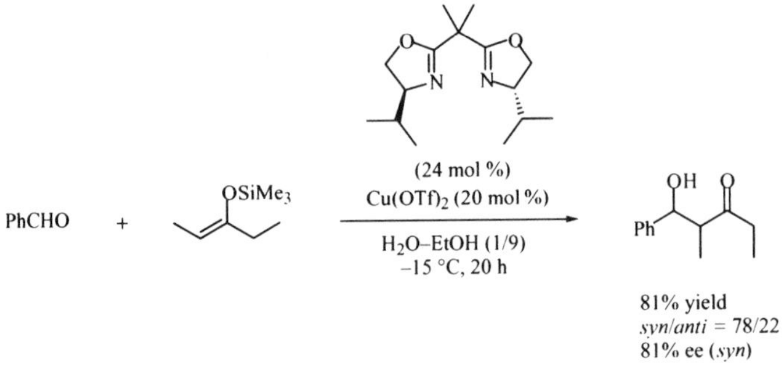

PhCHO + (structure: OSiMe₃ enol ether)

$$\xrightarrow[\substack{H_2O–EtOH\ (1/9) \\ -15\ °C,\ 20\ h}]{\substack{(24\ mol\ \%) \\ Cu(OTf)_2\ (20\ mol\ \%)}}$$

(product structure) (Eq. 3)

81% yield
syn/anti = 78/22
81% ee (*syn*)

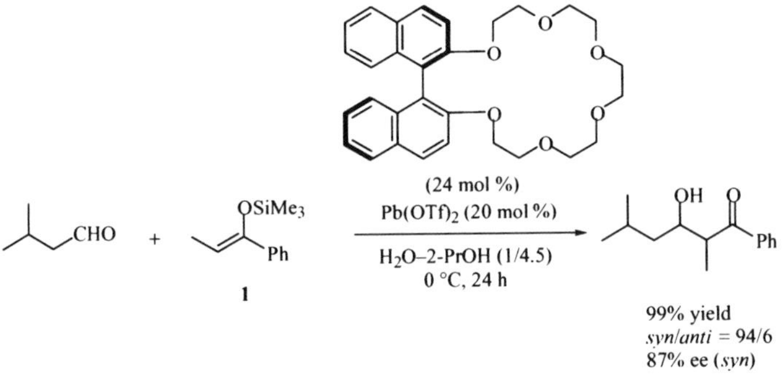

(structure: isovaleraldehyde) CHO + (structure: OSiMe₃ Ph) **1**

$$\xrightarrow[\substack{H_2O–2\text{-}PrOH\ (1/4.5) \\ 0\ °C,\ 24\ h}]{\substack{(24\ mol\ \%) \\ Pb(OTf)_2\ (20\ mol\ \%)}}$$

(product structure) (Eq. 4)

99% yield
syn/anti = 94/6
87% ee (*syn*)

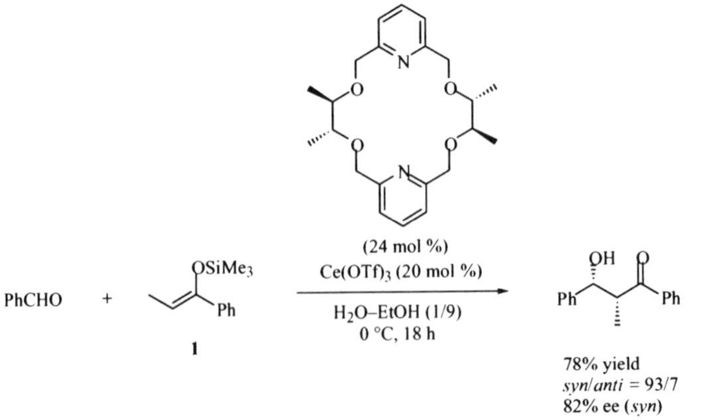

PhCHO + (structure: OSiMe₃ Ph) **1**

$$\xrightarrow[\substack{H_2O–EtOH\ (1/9) \\ 0\ °C,\ 18\ h}]{\substack{(24\ mol\ \%) \\ Ce(OTf)_3\ (20\ mol\ \%)}}$$

(product structure) (Eq. 5)

78% yield
syn/anti = 93/7
82% ee (*syn*)

factor to attain high selectivity. A study on the reaction profiles of the $Pr(OTf)_3$-catalyzed aldol reaction indicates that the crown ether does not significantly reduce the activity of $Pr(OTf)_3$. This retention of the activity in the presence of the crown ether is the key to realize the asymmetric induction in the present $Ln(OTf)_3$-catalyzed aldol reactions. The X-ray structure of $[Pr(NO_3)_2 \cdot 1]_3[Pr(NO_3)_6]$ shows that all of the methyl groups at the asymmetric carbon atoms of the crown ether are located in the axial positions. This conformation of the crown ring should be crucial to create an effective chiral environment around the Pr cation. This work will provide a useful concept for the design of chiral catalysts which function effectively in aqueous media.

Lewis Acids with Surfactant Molecules

While the Lewis acid-catalyzed aldol reactions in aqueous solvents described above were smoothly catalyzed by several metal salts, a certain amount of organic solvent such as THF had to be still combined with water to promote the reactions efficiently. To avoid the use of the organic solvents, we have developed a new reaction system in which metal triflates catalyze aldol reactions in water without using any organic solvents with the aid of a small amount of a surfactant such as sodium dodecyl sulfate (SDS).

The surfactant-aided Lewis acid catalysis was first found in the model reaction shown in Table I.(20) While the reaction proceeded sluggishly in the presence of 20 mol % $Yb(OTf)_3$ in water, remarkable enhancement of the reactivity was observed when the reaction was carried out in the presence of 20 mol % $Yb(OTf)_3$ in an aqueous solution of SDS (20 mol %, 35 mM), and the corresponding aldol adduct was obtained in 50% yield. The yield was improved when $Sc(OTf)_3$ was used as a Lewis acid catalyst. It was found that the surfactants influenced the yield, and that Triton X-100, a nonionic surfactant, was effective in the aldol reaction (but required long reaction time), while only a trace amount of the adduct was detected when using a representative cationic surfactant, cetyltrimethylammonium bromide (CTAB).

The results mentioned above prompted us to synthesize a more simplified catalyst, scandium tris(dodecyl sulfate) ($Sc(DS)_3$, Chart 1).(21) This new type of catalyst, "Lewis acid–surfactant-combined catalyst (LASC)", was expected to act *both* as a Lewis acid to activate the substrate molecules *and* as a surfactant to form dispersions in water.

Indeed, LASCs $Sc(DS)_3$ and 3 were found to catalyze the aldol reaction of benzaldehyde with silyl enol ether 1 in water. While $Sc(DS)_3$ and 3 are only slightly soluble in water, stable colloidal dispersions were formed upon addition of the aldehyde with stirring or vigorous mixing. Addition of 1, followed by stirring at room temperature for 4 h, gave the desired aldol adduct in high yields as shown in Eq. 6. It should be noted that hydrolysis of the silyl enol ether is not a severe problem under the reaction conditions in spite of the water-labile nature of silyl enol ethers under acidic conditions.

Table I. Effect of M(OTf)$_3$ and Surfactants

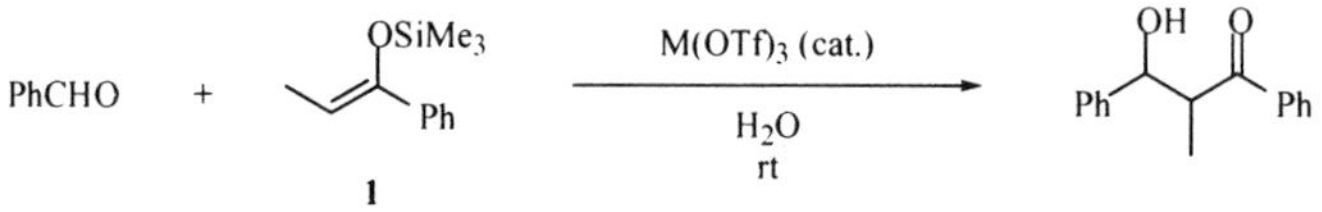

M(OTf)$_3$/mol %	Surfactant/mol %	Time/h	Yield/%
Yb(OTf)$_3$/20	—	48	17
Yb(OTf)$_3$/20	SDS/20	48	50
Sc(OTf)$_3$/10	SDS/20	4	88
Sc(OTf)$_3$/10	Triton X-100/20	60	89
Sc(OTf)$_3$/10	CTAB/20	4	trace

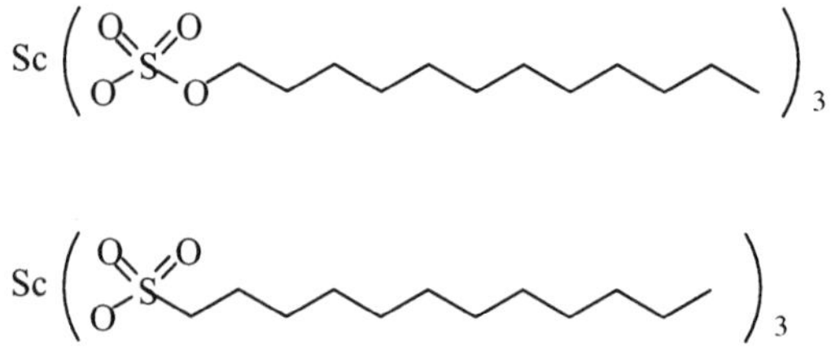

Chart 1

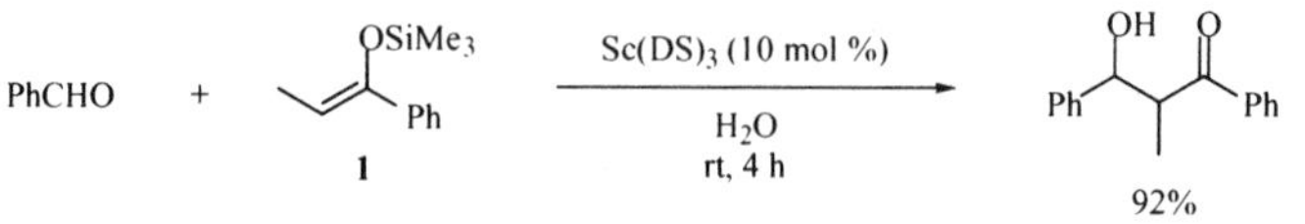

We also found that Sc(DS)$_3$ worked well in water rather than in organic solvents. For example, the initial rate of the aldol reaction in water (2.61 x 10^{-5}

M s^{-1}) was found to be 1.3 x 10^2 times higher than that in CH$_2$Cl$_2$ (1.95 x 10^{-7} M s^{-1}). Under neat conditions without using any solvents, the aldol reaction of benzaldehyde with **1** was also accelerated by Sc(DS)$_3$. However, the reaction was slower than in water and resulted in low yield (31%) due to formation of many by-products, showing the advantage of the use of water for the present reaction.

Various substrates have been successfully used in the present LASC-catalyzed aldol reaction. Aromatic as well as aliphatic, α,β-unsaturated, and heterocyclic aldehydes worked well. As for silyl enolates, silyl enol ethers derived from ketones as well as ketene silyl acetals derived from a thioester and an ester reacted well to give the corresponding adducts in high yields. It is noted that highly water-sensitive ketene silyl acetals reacted smoothly in water under these conditions.

In the work-up procedure for the aldol reactions stated above, the crude products were extracted with ethyl acetate after quenching the reactions. The addition of ethyl acetate in this procedure facilitates the phase separation between the organic and aqueous phase and makes the separation of organic products facile. It is more desirable, however, to develop a work-up procedure without using any organic solvents such as ethyl acetate. In addition, the development of a protocol for recovery and reuse of the catalysts is indispensable to apply the LASC system to large-scale syntheses. Therefore, it is worthy to mention that centrifugation of the reaction mixture of a LASC-catalyzed aldol reaction led to phase separation without addition of organic solvents. After centrifugation at 3500 rpm for 20 min, the colloidal mixture became a tri-phasic system where Sc(DS)$_3$ was deposited between a transparent water phase and an organic product phase. It is noted that this procedure enables, in principle, the recovery and reuse of LASCs and the separation of the organic products without using organic solvents.

In the LASC-catalyzed reactions, the formation of stable colloidal dispersions seemed to be essential for the efficient catalysis. We, thus, undertook the observation of the dispersions by means of several tools. Light microscopic observations of the colloidal dispersions revealed the formation of spherical colloidal particles in water (Figure 1a). The average sizes of the colloidal particles formed from **3** in the presence of benzaldehyde (**3**:PhCHO = 1:20) in water were measured by dynamic light scattering, and proved to be 1.1 µm in diameter. The shape and the size of the colloidal particles were also confirmed by transmission electron microscopy (Figure 1b).

From the observations stated above, we speculate the mechanism of the LASC-catalyzed aldol reactions in water as follows. In the presence of organic substrates, LASC molecules form stable colloidal particles in which the surfactant moiety of the LASCs surrounds the organic substrates and the countercations are attracted to the surface of the particles through electrostatic interactions between the anionic surfactant molecules and the cations. Although each Sc(III) cation is hydrated by several water molecules, they can be readily replaced by a substrate because of the high exchange rate of Sc(III) for substitution of inner-sphere water ligands. The substrates to be activated move to the interface from the organic phase, coordinate to the cations, and then react with nucleophilic substances there. In this mechanism, the roles of water in the reactions are assumed as follows: (i) hydrophobic interactions in water lead to increase of the activity coefficient of the catalyst and the substrates, resulting in the higher reaction rate in water; (ii) aggregation of the substrates through the hydrophobic interactions in water results in protection of water-labile substrates such as silyl enol ethers from their hydrolysis; (iii) hydration of Sc(III) ion and the counteranion by water molecules leads to dissociation of the LASC salt to form highly Lewis acidic species such as $[Sc(H_2O)_n]^{3+}$; (iv) rapid hydrolysis of initially formed scandium aldolates with water makes the catalytic turnover step in the aldol reactions fast. This may be a reason for the clean reaction in water compared with the reaction under neat conditions.

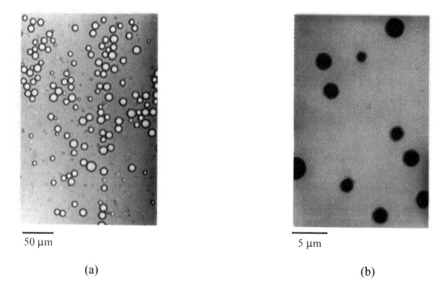

50 µm

5 µm

(a)   (b)

Figure 1. (a) Light micrograph of a mixture of 3 and PhCHO. (b) Transmission electron micrograph of a mixture of 3 and PhCHO.

Sc(DS)$_3$ was also found to be applicable to other Lewis acid-catalyzed reactions in water. For example, three-component Mannich-type reactions of aldehydes, amines, and silyl enolates in water were catalyzed by Sc(DS)$_3$ (Eq. 7).(*21b*) Mannich and related reactions provide a basic and useful method for the synthesis of β-amino carbonyl compounds, which constitute various pharmaceuticals, natural products, and versatile synthetic intermediates. It is noteworthy in the Sc(DS)$_3$-catalyzed Mannich-type reactions that the intermediate imines were formed *in situ* from aldehydes and amines even in the presence of water as a solvent.

$$\text{PhCHO} + \underset{\text{NH}_2}{\overset{\text{OMe}}{\bigcirc}} + \underset{\text{OMe}}{\overset{\text{OSiMe}_3}{\diagup}} \xrightarrow[\text{H}_2\text{O, rt, 5 h}]{\text{Sc(DS)}_3 \ (5 \text{ mol } \%)} \quad (\text{Eq. 7})$$

72%

Use of a phosphite ester as a nucleophilic component enables Sc(DS)$_3$-catalyzed three-component α-amino phosphonate synthesis in water (Eq. 8).(*22*) This reaction system provides a novel method for the synthesis of biologically important α-amino phosphonates.

$$\text{PhCH}_2\text{CH}_2\text{CHO} + \text{Ph}_2\text{CHNH}_2 + \text{P(OEt)}_3 \xrightarrow[\substack{\text{H}_2\text{O, 30 °C} \\ 1 \text{ h}}]{\substack{\text{Sc(DS)}_3 \\ (10 \text{ mol } \%)}} \quad (\text{Eq. 8})$$

95%

Furthermore, Sc(DS)$_3$ can also be applied to Michael reactions in water as shown in Eq 9.(*23*) As donor substrates, indoles also reacted with electron-deficient olefins smoothly (Eq. 10).(*24*) This is the first example of Lewis acid-catalyzed Friedel–Crafts-type reactions of aromatic compounds in water.

A catalytic asymmetric aldol reaction in water has been attained by a combination of copper bis(dodecyl sulfate) (Cu(DS)$_2$) and the chiral bis(oxazoline) ligand shown in Eq. 3.(*25*) In this case, addition of a Brønsted acid, especially a carboxylic acid such as lauric acid, is essential for good yield and enantioselectivity (Eq. 11). This is the first example of Lewis acid-catalyzed asymmetric aldol reactions in water without using organic solvents. Although the yields and the selectivities are still not yet optimized, it is noted that this enantioselectivity has been attained at ambient temperature in water.

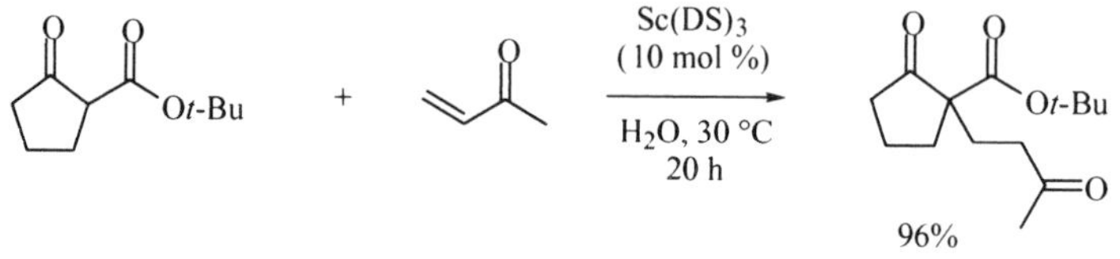

(Eq. 9)

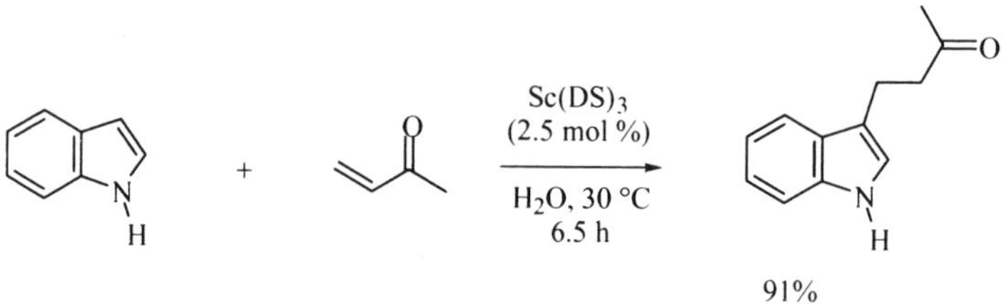

96%

(Eq. 10)

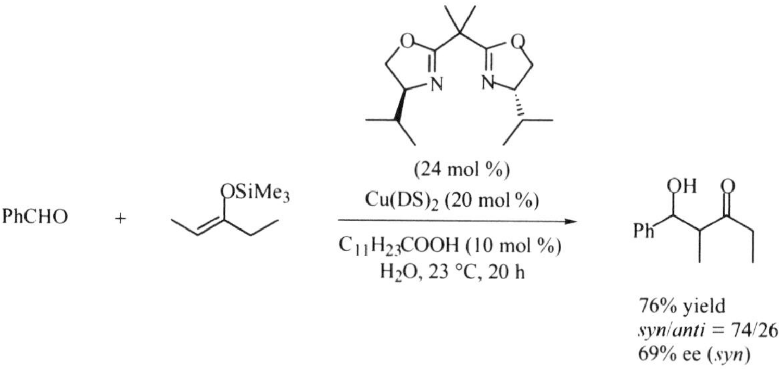

91%

(Eq. 11)

76% yield
syn/anti = 74/26
69% ee (*syn*)

The concept of surfactant-type catalysts described above was also found to be applicable to other catalytic systems such as Brønsted acid-catalyzed reactions (*26*) and palladium-catalyzed reactions.(*27*)

Lewis Acid Catalysis in Supercritical Carbon Dioxide

Although water works as a Lewis base to coordinate to the Lewis acid in the above reactions, the coordination occurs under equilibrium conditions and Lewis acid catalysis has been performed efficiently in such media. Similarly, it was expected that such Lewis acids would work well in supercritical carbon dioxide (scCO$_2$), which has also been regarded as a desirable substitute for some toxic organic solvents to accomplish benign chemical reactions.(*28*)

Diels–Alder reactions of carbonyl dienophiles with dienes (Eq. 12) and aza Diels–Alder reactions of imines with a diene (Eq. 13) were found to be successfully carried out using scandium tris(heptadecafluorooctanesulfonate) (Sc(O$_3$SC$_8$F$_{17}$)$_3$) as a Lewis acid catalyst in scCO$_2$.(*29*) It was revealed that the length of perfluorocarbon chains of the scandium catalyst was an essential factor for the catalytic activity in scCO$_2$.

Friedel–Crafts acylation in scCO$_2$ was also attained by scandium catalysts. Different from the Diels–Alder reactions mentioned above, scandium tris(nonafluorobutanesulfonate) (Sc(O$_3$SC$_4$F$_9$)$_3$) gave the best results.(*30*) The use of the water-compatible scandium catalyst as well as scCO$_2$ as a solvent is expected to lead to benign chemical processes.

$$\text{Sc(O}_3\text{SC}_8\text{F}_{17})_3 \text{ (1 mol \%)}, \quad \text{scCO}_2, \quad 50\ ^\circ\text{C, 150 atm, 24 h} \quad >99\% \qquad \text{(Eq. 12)}$$

$$\text{Sc(O}_3\text{SC}_8\text{F}_{17})_3 \text{ (5 mol \%)}, \quad \text{scCO}_2, \quad 40\ ^\circ\text{C, 100 atm, 1 h} \quad 86\% \qquad \text{(Eq. 13)}$$

$$\text{Sc(O}_3\text{SC}_4\text{F}_9)_3 \text{ (10 mol \%)}, \quad \text{scCO}_2, \quad 50\ ^\circ\text{C, 200 atm, 24 h} \quad 83\% \qquad \text{(Eq. 14)}$$

Conclusion

In this paper, we described Lewis acid catalysis in clean solvents, water and scCO$_2$. Various metal salts such as rare earth metal triflates can function as recoverable and reusable Lewis acids in aqueous media. Furthermore, LASCs promoted reactions in water without any organic solvents. Scandium catalysts are also successfully used in $scCO_2$. The Lewis acids described in this chapter are expected to be new types of catalysts providing some solutions for environmental problems. In addition, reactions in such unconventional media often exhibit unusual reactivity and selectivity, and, therefore, will lead to development of efficient catalytic systems which cannot be attained in conventional organic solvents.

Acknowledgment

Our work was partially supported by CREST, Japan Science and Technology Corporation (JST) and a Grant-in-Aid for Scientific Research from the Ministry of Education, Science, Sports, and Culture, Japan.

References

1. Schinzer, D., Ed. *Selectivities in Lewis Acid Promoted Reactions,* Kluwer Academic Publishers: Dordrecht, 1989.
2. (a) *Organic Synthesis in Water;* Grieco, P. A., Ed.; Blacky Academic and Professional: London, 1998. (b) Li, C.-J. *Chem. Rev.* **1993**, *93*, 2023.
3. (a) Kobayashi, S. *Lanthanides: Chemistry and Use in Organic Synthesis;* Springer: Heidelberg, 1999, p.63. (b) Kobayashi, S. *Eur. J. Org. Chem.* **1999**, 15. (c) Kobayashi, S. *Synlett* **1994**, 689.
4. (a) Mukaiyama, T.; Narasaka, K.; Banno, T. *Chem Lett.* **1973**, 1011. (b) Mukaiyama, T.; Banno, K.; Narasaka, K. *J. Am. Chem. Soc.* **1974**, *96*, 7503.
5. Mukaiyama, T. *Org. React.* **1982**, *28*, 203.
6. Chan, T.-H. In *Comprehensive Organic Synthesis;* Trost, B. M. Ed.; Pergamon Press: New York, 1991; Vol. 2, Chapter 2.3.
7. (a) Baes, Jr. C. F.; Mesmer, R. E. *The Hydrolysis of Cations,* John Wiley & Sons: New York, 1976, p. 129. (b) Yatsimirksii, K. B.; Vasil'ev, V. P. *Instability Constants of Complex Compounds*; Pergamon: New York, 1960. (c) Martell, A. E., Ed.;*Coordination Chemistry*; ACS Monograph 168; American Chemical Society: Washington, DC, 1978; Vol. 2.
8. (a) Thom, K. F., US Patent 3615169 (1971); CA 1972, 76, 5436a. (b) Forsberg, J. H.; Spaziano, V. T.; Balasubramanian, T. M.; Liu, G. K.; Kinsley, S. A.; Duckworth, C. A.; Poteruca, J. J.; Brown, P. S.; Miller, J. L. *J. Org. Chem.* **1987**, *52*, 1017.
9. Kobayashi, S. *Chem. Lett.* **1991**, 2187.

10. (a) Kobayashi, S.; Hachiya, I. *Tetrahedron Lett.* **1992**, 1625. (b) Kobayashi, S.; Hachiya, I. *J. Org. Chem.* **1994**, *59*, 3590.
11. Haggin, J. *Chem. Eng. News* **1994**, Apr 18, 22.
12. Kobayashi, S.; Nagayama, S.; Busujima, T. *J. Am. Chem. Soc.* **1998**, *120*, 8287.
13. Hachiya, I.; Kobayashi, S. *J. Org. Chem.* **1993**, *58*, 6958.
14. Kobayashi, S.; Ishitani, H. *J. Chem. Soc. Chem. Commun.* **1995**, 1379.
15. (a) Carreira, E. M. in *Comprehensive Asymmetric Catalysis*; Jacobsen, E. N.; Pfaltz, A.; Yamamoto, H. Eds.; Springer-Verlag: Berlin, 1999, pp 997. (b) Machajewski, T. D.; Wong, C.-H. *Angew. Chem. Int. Ed.* **2000**, *39*, 1352.
16. (a) Mukaiyama, T.; Kobayashi, S.; Uchiro, H.; Shiina, I. *Chem. Lett.*, **1990**, 129. (b) Kobayashi, S.; Fujishita, Y.; Mukaiyama, T. *Chem. Lett.*, **1990**, 1455.
17. (a) Kobayashi, S.; Nagayama, S.; Busujima, T. *Chem. Lett.* **1999**, 71. (b) Kobayashi, S.; Nagayama, S.; Busujima, T. *Tetrahedron* **1999**, *55*, 8739.
18. Nagayama, S.; Kobayashi, S. *J. Am. Chem. Soc.* **2000**, *122*, 11531.
19. Kobayashi, S.; Hamada, T.; Nagayama, S.; Manabe, K. *Org. Lett.* in press.
20. Kobayashi, S.; Wakabayashi, T.; Nagayama, S.; Oyamada, H. *Tetrahedron Lett.* **1997**, *38*, 4559.
21. (a) Kobayashi, S.; Wakabayashi, T. *Tetrahedron Lett.* **1998**, *39*, 5389. (b) Manabe, K.; Mori, Y.; Wakabayashi, T.; Nagayama, S.; Kobayashi, S. *J. Am. Chem. Soc.* **2000**, *122*, 7202.
22. Manabe, K.; Kobayashi, S. *Chem. Commun.* **2000**, 669.
23. Mori, Y.; Kakumoto, K.; Manabe, K.; Kobayashi, S. *Tetrahedron Lett.* **2000**, *41*, 3107.
24. Manabe, K.; Aoyama, N.; Kobayashi, S. *Advanced Synth. Catal.* in press.
25. Kobayashi, S.; Mori, Y.; Nagayama, S.; Manabe, K. *Green Chemistry* **1999**, *1*, 175.
26. (a) Manabe, K.; Mori, Y.; Kobayashi, S. *Synlett* **1999**, 1401. (b) Manabe, K.; Kobayashi, S. *Org. Lett.* **1999**, *1*, 1965. (c) Manabe, K.; Mori, Y.; Kobayashi, S. *Tetrahedron* in press.
27. Kobayashi, S.; Lam, W. W.-L.; Manabe, K. *Tetrahedron Lett.* **2000**, *41*, 6115.
28. Jessop, P. G.; Ikariya, T.; Noyori, R. *Chem. Rev.* **1999**, *99*, 475.
29. Matsuo, J.; Tsuchiya, T.; Odashima, K.; Kobayashi, S. *Chem. Lett.* **2000**, 178.
30. Kawada, A.; Mitamura, S.; Matsuo, J.; Tsuchiya, T.; Kobayashi, S. *Bull. Chem. Soc. Jpn.* **2000**, *73*, 2325.

Chapter 12

Developing Green Chemistry: Organometallic Reactions in Aqueous Media

Tak Hang Chan, Lianhai Li, Yang Yang, and Wenshuo Lu

Department of Chemistry, McGill University, 801 Sherbrooke Street West, Montreal, Quebec H3A 2K6, Canada

Abstract: Various issues which affect organometallic reactions in aqueous media have been considered. These include the activation of metals, the allylation of carbonyl compounds with different metals, the regio- and stereo-selectivity of the allylation reactions; and the mechanisms of these aqueous organometallic reactions.

Introduction

A critical question in the use of clean solvents such as water, supercritical carbon dioxide or ionic liquids for chemical synthesis is: do these solvents react with the reagents? This question is particularly relevant in the case of reactive organometallic reagents including the Grignard and the organolithium compounds. These organometallic reagents are very useful in chemical synthesis, but they are well known to react with water or carbon dioxide. They also undergo acid-base reactions readily with quarternary ammonium, sulfonium or phosphonium salts, structures which are commonly used for ionic liquids. The reactions of the clean solvents with these organometallic reagents tend to limit the usefulness of these solvents for chemical synthesis. In order to overcome this critical problem, alternative organometallic reagents and reactions which are compatible with water, supercritical carbon dioxide or ionic liquids must be developed for the necessary synthetic transformations. This chapter describes our research in this area, with emphasis on organometallic reactions in aqueous media, and touches on the various issues which need to be considered.

The allylation of carbonyl compounds is a very important reaction in organic synthesis. In the past decade, we and others have discovered that this transformation could be achieved in aqueous meida through a Barbier type reaction according to Scheme 1. Metals such as In (1), Sn (2), Zn (3), Mg (4), Bi (5), Sb (6), Pb (7), and Mn (8) have all been found to mediate the coupling between allyl halides and carbonyl compounds to give the corresponding homoallylic alcohols. Among these metals, indium is generally considered to be the metal of choice for the reaction (9). Furthermore, we have demonstrated recently that the indium-mediated reaction proceeds through an allylindium(I) intermediate (10). Other metals give more side products (4, 8) , or require acidic conditions or sonication (3, 4), or need to be generated in situ in a reduction step (5, 6) in organic solvents.

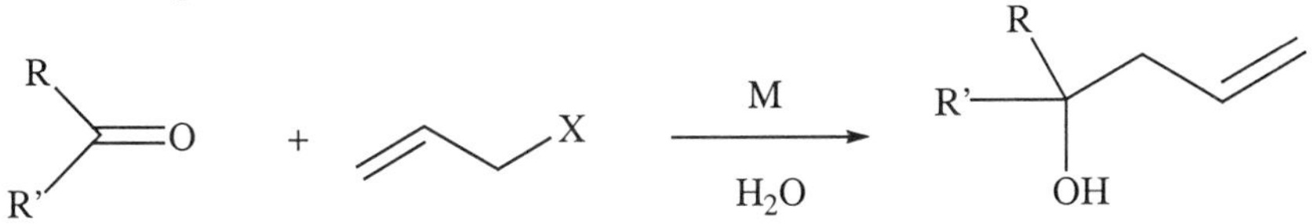

Scheme 1

We are interested in the following issue. In order to make organometallic reactions in aqueous media truly useful for organic synthesis, we must have a stable of diverse organometallic reagents with different chemo-, regio- and stereoselectivity. Can some of the other metals be as versatile as indium and would they show different selectivities from the indium reaction?

Metal Activation in Aqueous Media

Other than the alkali and alkaline earth metals, most metals are unreactive in water. For many metals, this is not because of the lack of inherent reactivity, but because they can readily form oxides which are sparingly soluble in water and prevent them from further reactions. The use of aluminum or pewter (a tin and antimony alloy) for food utensils is a good illustration of such a phenomenum. In the antimony-mediated allylation reaction (Scheme 1, M=Sb), commercially available antimony metal powder was not successful in mediating the reaction, presumably because of the formation of the insoluble oxide on the metal surface. The antimony metal must be generated in situ through the reduction of antimony trichloride with a reducing agent (6, 11) in an organic solvent (or mixed aqueous organic solvent) and used immediately. This renders the use of antimony less convenient. We recently found that aqueous soluble fluoride salts are quite effective in "activating" antimony in aqueous media to mediate the allylation reaction (12). Whereas there was no formation of the homoallylic alcohol after a week of vigorous stirring of a mixture of antimony, benzaldehyde and allyl bromide in distilled water, a good yield of the product was obtained in 16 hrs when the same mixture was stirred in a 1M solution of RbF or CsF (Table 1). At 2M concentration, NaF and KF were found to be as effective as RbF and CsF. On the other hand, even at 2M concentration, KBr was completely ineffective, and KCl and KI were much less effective.

Table I. Effect of halide salts on the allylation of benzaldehyde with antimony (12)[a]

Entry	Metal halide	Yield %[b] (hr/d), 2M	Yield %[b] (hr), 1M
1	LiF	34 (16)	10 (16)
2	NaF	100 (16)	28 (16)
3	KF	100 (16)	53 (16)
4	RbF	100 (16)	89 (16)
5	CsF	100 (16)	92 (16)
6	KCl	62 (4d)	
7	KBr	0 (4d)	
8	KI	90 (4d)	

(a) Reaction conditions: benzaldehyde (1 mmol), allyl bromide (2.5 mmol) and antimony powder (2 mmol) in 4 mL solution at room temperature. (b) Yield was based on the ^{1}H NMR analysis of the crude product.

Using the 2M KF as the standard reaction conditions, commercial antimony metal was able to allylate a number of aldehydes to give excellent yields of the products according to Scheme 1. Interestingly, nitrobenzaldehyde was successfully allylated. Usually, the nitro group is sensitive to reduction. In this sense, the use of fluoride salts as activating agent is superior to the use of reducing agents reported previously for activating antimony.

The activation of metal in aqueous media by fluoride salts appears to have broad applicability, as we shall demonstrate in later parts of this chapter.

Organometallic Reagents in Aqueous Media

In organic synthesis using organic solvents, the classical Barbier reaction has been largely supplanted by the Grignard or similar reactions where the organometallic reagent is first generated. The reason is that in most reactions involving carbonyl compounds, the use of metal directly can often lead to side products such as pinacol coupling or reduction. Furthermore, the reactivity of the organometallic reagents can be modulated through the use of different metals or addition of ligands. This adds considerably to the versatility of organometallic reactions in synthesis. In the aqueous allylation reaction (Scheme 1) involving various metals, are discreet organometallic reagents generated? And if the answer is positive, do these organometallic reagents have finite lifetimes in the aqueous environment and can their reactivity be modulated? In 1999, we demonstrated that an allylindium (I) intermediate **1** is formed in the indium-mediated allylation of carbonyl compounds (10) (Scheme 2). The allylindium(I) (**1**) thus generated could be observed by NMR and has a finite lifetime in water.

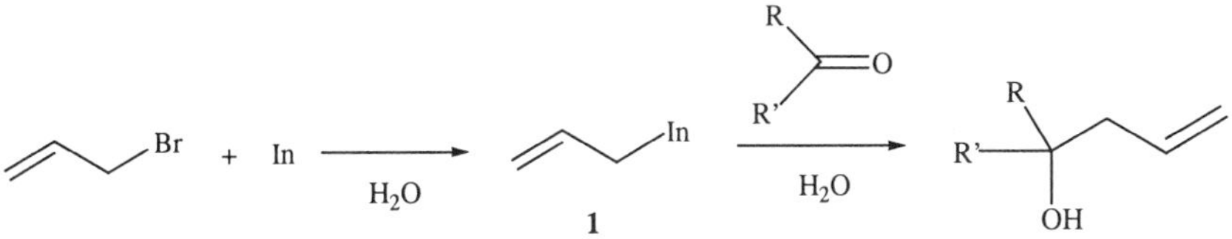

Scheme 2

Similarly, in the tin-mediated allylation reaction, allyltin intermediates are generated (13). Both allyltin(II) bromide (**2**) and diallyltin(IV) dibromide (**3**) are formed, and can be observed by NMR in the aqueous media (Scheme 3).

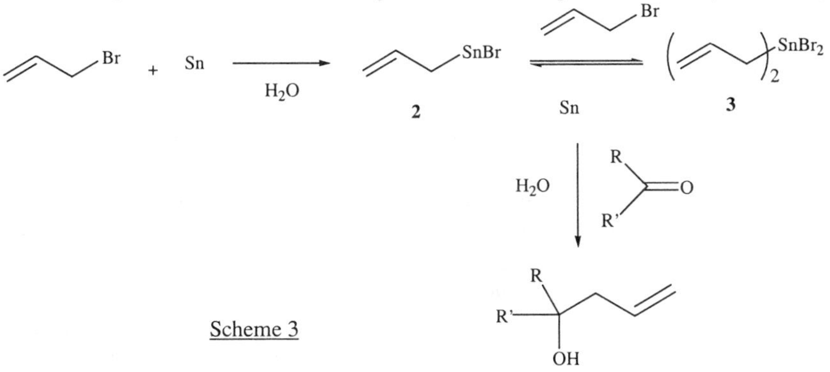

Scheme 3

We have recently showed that similar allylstibine intermediates **4** and **5** can be formed in the antimony-mediated allylation conditions according to Scheme 4.

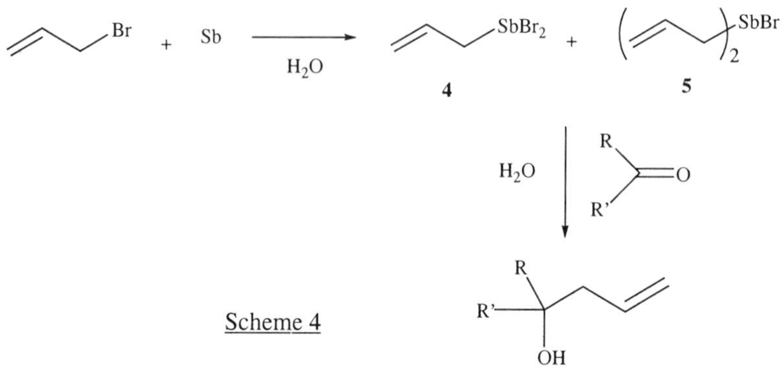

Scheme 4

H_2O 4   5

H_2O

Even though these discreet but transient organometallic intermediates (**1-5**) will eventually hydrolyse in the aqueous media, we do know that they have sufficient lifetime for them to be observed by ^{1}H NMR. Furthermore, it has been observed that the allyltin intermediates **2** and **3** can survive in water for more than a day, whereas allylindium(I) (**1**) and allylstibines **4** and **5** disappear in water in an hour.

On the other hand, in the magnesium and zinc-mediated allylation reactions, if organometallic intermediates were formed, they could not be observed in the aqueous media by ^{1}H NMR. This is not surprising in view of the known reactivity of the Grignard and organozinc reagents with water. It remains to be seen whether for metals such as Bi, Pb or Mn, discreet organometallic intermediates can exist in water with finite lifetimes. As a comparison, allylmercury bromide is a well characterised compound, indefinitely stable in water and does not react with aldehydes or ketones (14).

Chemoselectivity

Organometallic reagents in organic solvents show different chemoselectivities depending on the metals involved. This ability to discriminate among different functional groups is extremely important in the use of organometallic reagents in organic synthesis. As we discover more organometallic reactions in aqueous media, it is equally important to understand their different chemoselectivities. In Table 2, we have summarized the knowledge obtained thus far concerning the chemoselectivities of organo-indiums, tin, antimony, bismuth and mercury reagents towards various carbonyl functions. In general, all these reagents, with the possible exception of allylmercury bromide (14), react with aldehydes in aqueous media to give the addition products. Allylindium **1** and diallyltin dibromide (**3**) react with ketones readily, but allylantimony compounds **4** and **5** do not. The chemoselectivity can be quite subtle. For example, bismuth, under the fluoride salt activation conditions in 2M KF aqueous solution, allylates

Table 2. Chemoselectivity of various organometallic reagents in aqueous media

	Allyl-indium (I) **1**	CpIn (**6**)	Allyltin **3**	Allylsti-bines **4** and **5**	Allyl-bromide/ bismuth[a]	Allyl-mercury bromide
aldehyde	yes	yes	yes	yes	yes	no[c]
ketone	yes	no	yes	no	yes/no[b]	no
ester	no	no	no	no	no	no
acylpy-razole	yes	-	-	-	-	-
Carbox-ylic acid	no	-	no	-	-	-

(a) An allylbismuth intermediate is presumed. (b) Reacts with cyclohexanone, but not with acetophenone. (c) Reacts with aldehydes on activation with tetrahexylammonium bromide (14).

cyclohexanone, but not open chain aliphatic ketones or acetophenone (15). The reactivity of the organometallic reagent is modulated not only by the metal, but by the structure of the organic group as well. Cyclopentadienylindium(I) (**6**) reacts readily with aldehydes but less so with ketones (16). These aqueous "stable" organometallic reagents do not react with esters nor with carboxylic acids. Active esters such as acylimidazoles and acylpyrazoles react with allylindium **1** (17) in aqueous media, but their reactions with the other reagents have not been investigated. These results suggest that different chemoselectivity can indeed be achieved in aqueous media with different organometallic reagents.

Regioselectivity

Our knowledge about the influence of metal on the regioselection of the aqueous metal-mediated allylation reaction can be summarized in Scheme 5. For allyl halides with a bulky substituent (R'=t-butyl or trimethylsilyl), the indium-mediated reaction gives the γ-adduct **7** (18). However, for most unsymmetrically substituted allylic halides (R'=alkyl, aryl, esters or halogens), the reaction has the carbon-carbon bond formation at the more substituted carbon to give the α-adduct **8**. The regioselection appears to be the same for all the metals examined thus far (M=In, Zn, Sn, Sb and Bi). There is a need to find metals which can mediate the allylation with opposite regioselectivity.

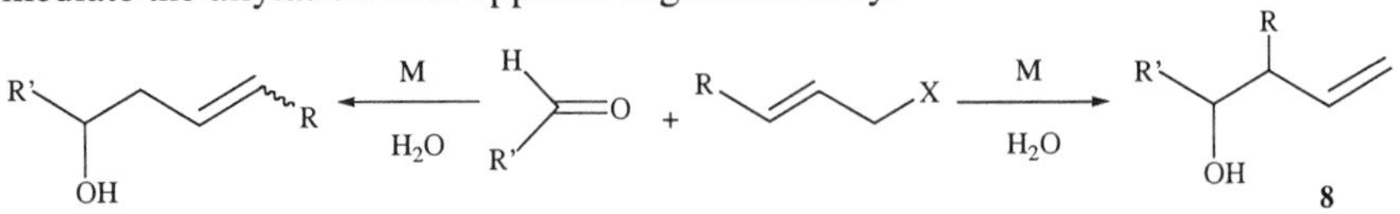

7 R=t-Bu, TMS   M=In

M=In, Zn, Sn, Sb, Bi

R=Me, Ph, CO$_2$R", Br, Cl

Scheme 5

On the other hand, the choice of metal has an effect on the regioselectivity of the metal-mediated coupling of propargyl halides with carbonyl compounds in aqueous media. The role of different metals on the ratio of **9** and **10** in the coupling of propargyl bromide with benzaldehyde and n-heptanal (Scheme 6, R=Ph or n-C$_6$H$_{13}$) has been investigated (19). In general, In, Sn, and Zn give preferentially the propargyl compound **9** as the major product, whereas Bi and Cd give both **9** and **10** in nearly equal proportions.

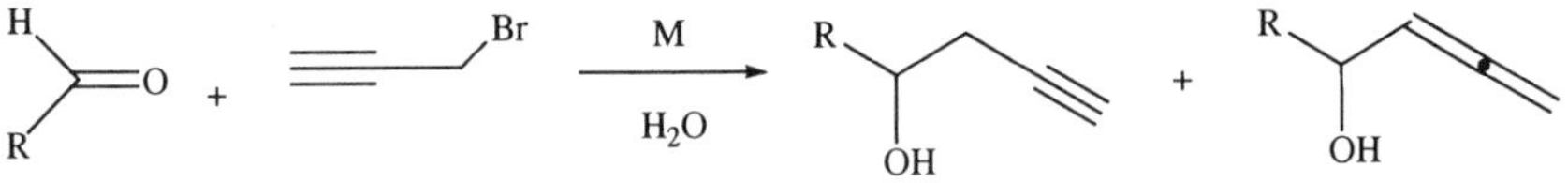

Scheme 6

For γ-substituted propargyl bromides, the results are somewhat different. In the indium-mediated coupling reaction in aqueous media, the allenylic compound **12** is the major product obtained (Scheme 7, Y=alkyl, aryl and silyl) (20). Similar reactions with other metals have not been performed to see if the regioselectivity may be changed. There is a strong possibility, however, that the zinc-mediated reaction may show a different regioselectivity (21[21]). This is based on our observation that in the coupling of γ-substituted propargyl bromides with the sulfonimines of

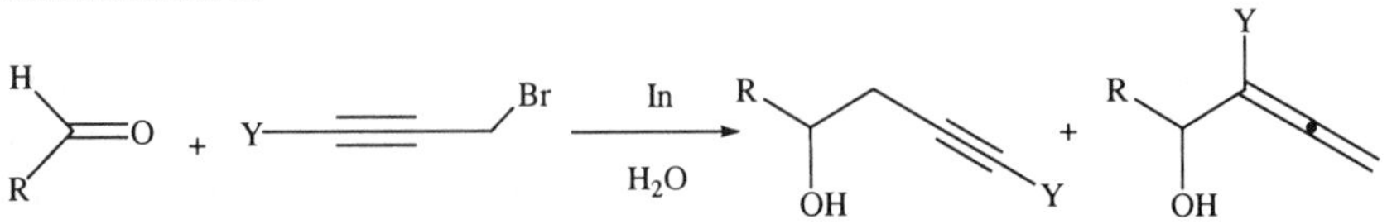

Y=alkyl, aryl, silyl   **11** minor   **12** major

Scheme7

arylaldehydes mediated by zinc in aqueous ammonium chloride solution, the homopropargyl adduct **13** is the only observed regioisomeric product (Scheme 8, Y=Me or Ph) (22). Similar reaction with indium does not lead to the formation of **13** because the sulfonimine is hydrolyzed prior to coupling. However, in THF, a mixture of allenylic and homopropargylic isomers is obtained.

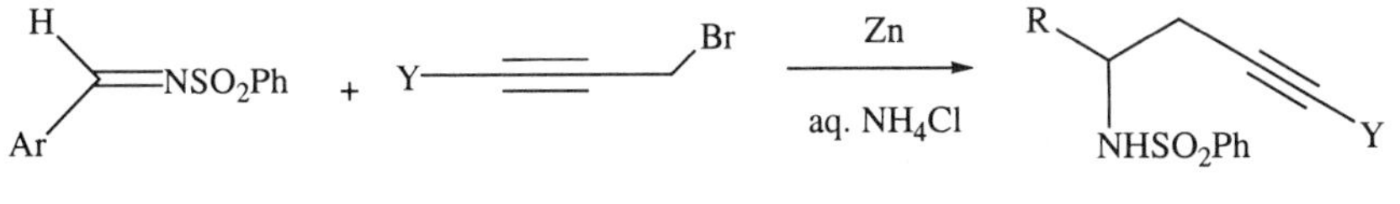

Scheme 8

Diastereoselectivity

The diastereoselectivity of the indium-mediated allylation reaction in water has been reviewed recently (9). Here, we are interested in the change in diastereoselectivity with different organometallic reagents. In general, the limited data available for zinc or tin or antimony suggest that they are similar to indium, giving preferentially the anti-adduct **14** over the syn-adduct **15** in the crotylation of aldehydes (Scheme 9). Interestingly, bismuth seems to give

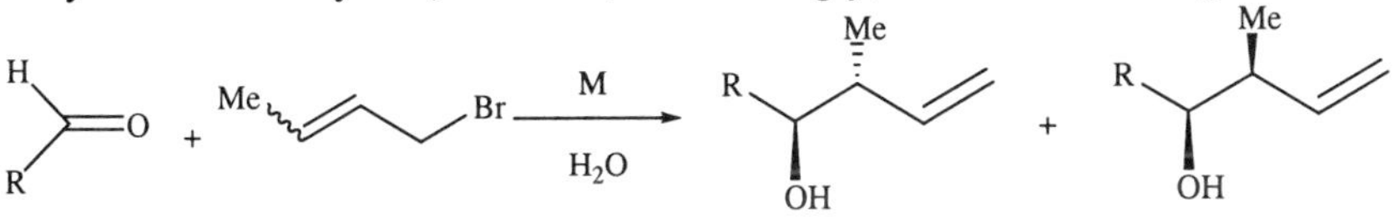

Scheme 9

preferentially the syn-adduct **15** under fluoride activation conditions (15). The data are summarized in Table 3. While the diastereoselectivity is not high, the results suggest that modulation of selectivity is possible with different metals. Further exploration is clearly warranted.

Aluminum/Fluoride Salt-Mediated Pinacol Coupling and Reduction of Carbonyl Compounds in Aqueous Media.

The coupling of carbonyl compounds to give pinacols **16** and **17** in aqueous media has previously been achieved using Zn-Cu (23) or low valent titanium

Table 3. Diastereoselectivity in the crotylation reaction according to Scheme 9 with indium or bismuth.[a]

R=	Metal=	Yield %	Anti- **14**/syn- **15**
Ph	In	90	50:50
i-Pr	In	88	84:16
t-Bu	In	87	80:20
Ph	Bi	94	18:82
2-Furyl	Bi	90	26:74
Cyclohexyl	Bi	89	47:53

(a) Reaction conditions: A mixture of aldehyde (1 mmol), allyl bromide (2 mmol) and bismuth powder (2 mmol) in 1M NH_4HF_2 aqueous solution (2 mL) was stirred for 4 hrs at room temperature.

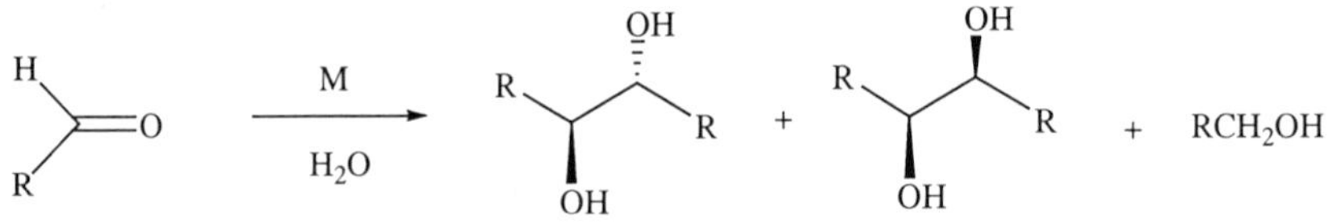

Scheme 10

reagents (24). The reaction is often accompanied by the reduction product **18** (Scheme 10).

We have chosen to examine the potential of aluminum metal for this reaction because of its low cost and ready availability. Aluminum has a first ionization potential of 5.986 eV which is lower than magnesium (7.65 eV) and comparable to lithium (5.39 eV). However, aluminum is resistant to water because it forms readily a thin film of aluminum oxide on the metal surface and shields itself from further reaction (25). Reports that trace amount of fluoride anion in water can accelerate the corrosion of aluminum metal (26) raises the possibility that fluoride anion may facilitate the removal of the insoluble aluminum oxide from the metal surface. Using benzaldehyde as the prototypal aldehyde, we examined its reaction with aluminum in aqueous media with various halide salts (27). The results are summarized in Table 4.

Table 4. Effects of metal halide salts on the reaction of benzaldehyde with Al in water (Scheme 10, R=Ph)[a]

Metal halide	Time (hr)	Conversion (%)	% 16	% 17	% 18
-	1 week	0	0	0	0
KF	16	100	41	45	13
KCl	16	0	0	0	0
KBr	16	0	0	0	0
KI	16	0	0	0	0
LiF	5 days	100	40	36	24
NaF	2 days	100	42	42	16
RbF	16	100	43	47	10
CsF	16	100	46	46	8
Bu_4NF	16	100	40	59	1
FeF_2	16	100	0	0	100
$FeCl_2$	16	0	0	0	0
CoF_2	16	100	0	0	100
NiF_2	16	100	0	0	100
$NiBr_2$	16	0	0	0	0

(a) Reaction conditions: Metal halide (5 mmol) was dissolved or suspended in distilled water (5 mL) and benaldehyde (1 mmol) was added. To the mixture was added Al powder (2.5 mmol) in one portion. The mixture was stirred at room temperature for the time indicated.

It is clear that fluoride salts have a special activating effect on aluminum metal in aqueous media. In the absence of fluoride salt, Al/H_2O is completely inactive. Salts of chloride, bromide or iodide have no activating effect either. Furthermore, depending on the particular fluoride salt used, the reaction can show high chemoselectivity. Alkaline metal or tetrabutylammonium fluoride salts lead preferentially to the pinacols **16** and **17**, whereas the fluoride salts of Fe^{2+}, Co^{2+} and Ni^{2+} are very selective in promoting the reduction of benzaldehyde.

Interestingly, copper(II) fluoride promotes the aluminum-mediated pinacol coupling of arylaldehydes stereoselectively. The meso- compound **16** is fomed predominantly (Table 5). While there have been previous reports in the literature (28) of stereoselective coupling of carbonyl compounds to the dl-isomer **17**, this is the first example of selective formation of meso-pinacols under metal reduction conditions.

176

Table 5. Reaction of benzaldehydes with Al/CuF$_2$ in water (Scheme 10)[a]

Aldehydes	Yield by ^{1}H NMR (%) 16+17 (16:17)	Yield by ^{1}H NMR (%) 18	Isolated yield (%) 16+17 (16:17)
PhCHO	96 (16:1)	4	95 (35:1)
p-ClPhCHO	57 (15:1)	18	54 (24:1)
p-CF$_3$PhCHO	85 (3:1)	15	84 (3:1)
p-MePhCHO	83 (15:1)	17	82 (21:1)

(a) Reaction conditions: under an Ar atmosphere, Al (2.5 mmol, 99.95%) and CuF$_2$ (1 mmol) were stirred in distilled water (5 mL) for 40 min; then another portion of Al (2.5 mmol) was added and followed by aldehyde (1 mmol) immediately. Reaction was worked up in 6 h.

Conclusions

It is clear by now that many organometallic reactions can be carried out in aqueous media. Many of these reactions proceed through discreet organometallic intermediates which have finite lifetimes in the aqueous environment. The chemoselectivity, regioselectivity and the diastereoselectivity of these reactions can be modulated with the choice of different metals. Furthermore, using fluoride salts as the "activating" agent, many metals which are previously thought to be unreactive in water can now be used to mediate reactions. The expansion of the choice of metals, and the use of different fluoride salts, can greatly expand the scope and the selectivities of organometallic reactions in aqueous media.

Acknowledgment

Financial support of this research by NSERC of Canada is gratefully acknowledged.

References

(1) Li, C. J.; Chan, T. H. Tetrahedron Lett. **1991**, 32, 7017.
(2) Nokami, J.; Otera, J.; Sudo, T.; Okawara, R. Organometallics, **1983**, 2, 191.
(3) Petrier, C.; Luche, J. L. J. Org. Chem. **1983**, 50, 910.
(4) Zhang, W. C.; Li, C. J. J. Org. Chem. **1999**, 64, 3230.
(5) Wada, M.; Ohki, H.; Akiba, K. Y. Bull. Chem. Soc. Jpn. **1990**, 63, 2751.

(6) Wang, W.; Shi, L.; Huang, Y. *Tetrahedron* **1990**, *46*, 3315.

(7) Zhou, J. Y.; Jia, Y.; Sun, G. F.; Wu, S. H. *Synth. Commun.* **1997**, *27*, 1899.

(8) Li, C. J.; Meng, Y.; Yi, X. H.; Ma, J. H.; Chan, T. H. *J. Org. Chem.* **1998**, *63*, 7498.

(9) For a recent review of the In reaction, see Li, C. J.; Chan, T. H. *Tetrahedron* **1999**, *55*, 11149.

(10) Chan, T. H.; Yang, Y. *J. Am. Chem. Soc.* **1999**, *121*, 3229.

(11) Ren, P. D.; Jin, Q. H.; Yao, Z. P. *Synth. Commun.* **1997**, *27*, 2761.

(12) Li, L.-H.; Chan, T. H. *Tetrahedron Lett.* **2000**, *41*, 5009.

(13) Chan, T. H.; Yang, Y.; Li, C. J. *J. Org. Chem.* **1999**, *64*, 4452.

(14) Chan, T. H.; Yang, Y. *Tetrahedron Lett.* **1999**, *40*, 3863.

(15) Li, L.-H.; Chan, T. H. results to be published.

(16) Yang, Y.; Chan, T. H. *J. Am. Chem. Soc.* **2000**, *122*, 402.

(17) Bryan, V. J.; Chan, T. H. *Tetrahedron Lett.* **1997**, *38*, 6493.

(18) Isaac, M. B.; Chan, T. H. *Tetrahedron Lett.* **1995**, *36*, 8957.

(19) Yi, X. H.; Meng, Y.; Hua, X. G.; Li, C. J. *J. Org. Chem.* **1998**, *63*, 7472.

(20) Isaac, M. B.; Chan, T. H. *J. Chem. Soc., Chem. Commun.* **1995**, 1003.

(21) Bieber, L. W.;da Silva, M. F.; da Costa, R. C.; silva, L. O. S. *Tetrahedron Lett.* **1998**, *39*, 3655. These authors briefly examined the zinc-mediated coupling of 1-bromo-2-butyne with benzaldehyde.

(22) Lu, W.; Chan, T. H. results to be published.

(23) Delair, P.; Luche, J. L *J. Chem. Soc., Chem. Commun.* **1989**, 398.

(24) Clerici, A.; Porta, O. *J. Org. Chem.* **1989**, *54*, 3872 and references cited therein.

(25) Pinacol coupling of carbonyl compounds with Al powder in methanol promoted by KOH or NaOH has been reported. See: Sahade, D. A.; Mataka, S.; Sawada, T.; Tsukinoki, T.; Tashiro, M. *Tetrahedron Lett.* **1997**, *38*, 3745.

(26) Tennakone, K.; Wickramanayake, S. *Nature* **1987**, *325*, 202.

(27) Li, L.-H.; Chan, T. H. *Org. Lett.* **2000**, *2*, 1129.

(28) Barden, M. C.; Schwartz, J. *J. Am. Chem. Soc.* **1996**, *118*, 5484 and references cited therein.

Chapter 13

Organic Reactions in Water and Other Alternative Media: Metal-Mediated Carbon–Carbon Bond Formations

Chao-Jun Li, John X. Haberman, Charlene C. K. Keh, Xiang-Hui Yi,
Yue Meng, Xiao-Gang Hua, Sripathy Venkatraman, Wen-Chun Zhang,
Tien Nguyen, Dong Wang, Taisheng Huang, and Jianheng Zhang

Department of Chemistry, Tulane University, New Orleans, LA 70118

Metal-mediated carbonyl additions were investigated using
several alternative media. Using the allylation as a model,
metals were examined across the periodic table in mediating
carbonyl additions in water. The effectiveness of metals as the
mediator for the reaction was found closely related to the
location of the metals in the periodic table and was tentatively
rationalized via inner-sphere and outer-sphere single-electron-
transfer processes. It was also found that, for allylation of
aldehydes, water, liquid carbon dioxide, and neat conditions
are all effective media when metals such as indium, tin and
zinc were used as the mediators. Various homoallyl alcohols
were prepared by these reactions. Under these conditions,
allyl halides were found much more reactive than other types
of organic halides. The cause for the special reactivity of allyl
halides as well as the mechanism for forming allylmetal
intermediates is discussed.

Recently, owing to an increased awareness of the detrimental effects that organic solvents have on the environment, a substantial amount of research has been devoted to exploring chemistry that is more environmentally friendly (1). A large part of this endeavor concerns itself with solvents. Almost all chemical processes make use of organic solvents at some point. These organic solvents, used in academic research and in industry, are often harmful to the environment, and as a result are frequently subject to government restrictions and high waste disposal costs. Consequently, methods that successfully minimize their use are the focus of much attention. The metal-mediated reaction of carbonyl compounds with organic halides is an important method in creating a carbon-carbon bond. Such reactions are widely used in chemical syntheses. Various metals, such as zinc (2), magnesium (3), lithium (4), and metalloids have been used to tailor the reactivity to control chemo-, regio-, diastereo-, and enantioselectivities (Eq. 1). Conventionally, such reactions are carried out in dry organic solvent such as tetrahydrofuran and ether. Other solvents used for such reactions include benzene, dichloromethane, chloroform, toluene, N,N-dimethylformate. With the impulse of developing cleaner and safer alternative solvents for chemical syntheses, we have investigated metal-mediated carbonyl additions in water, liquid carbon dioxide, and under neat conditions.

$$
\underset{R}{\overset{O}{\underset{H}{\bigtriangleup}}} \; + \; R'X \; \xrightarrow{M} \; \underset{R}{\overset{OH}{\underset{R'}{\bigtriangleup}}} \qquad (1)
$$

Metal-Mediated Carbonyl Additions In Water

Water is generally considered undesirable for organometallic reactions. Since the 1980s, however, it has been found that metal-mediated reactions could in fact be carried out in water. Several metals including tin (5); zinc (6); indium (7), bismuth (8), etc., have been used for such purposes. This early development has been reviewed previously (9). The advantages of using the aqueous medium reaction for synthetic applications have been exemplified in the synthesis of several natural products. However, most successes in this area are confined to allylation reactions. Similar reactions with other organic halides have not been as successful. One of our major recent research efforts is to understand the scope and limitation of the reaction and to find potential ways to overcome these limitations.

In order to understand the scope of the metal-mediated carbonyl addition reaction in water, a question that needs to be answered is the upper limit regarding metals to be used for this reaction. Towards this goal we have examined metals across the periodic table as well as referencing related works reported in the literature by using the allylation reaction of aldehydes as a model (Eq. 2). Some interesting findings are summarized as the following:

$$R \overset{O}{\underset{H}{\bigwedge}} + \diagdown\!\diagup\!\diagdown X \xrightarrow[H_2O]{M} R \overset{OH}{\bigwedge}\!\diagdown\!\diagup \quad (2)$$

The alkaline metals (**IA**) react violently upon contacting with water and are thus not suitable for the aqueous reaction. Alkaline earth metals such as organomagnesium compounds, especially the Grignard reagents, have played a key role in organic and organometallic synthesis for nearly 100 years (10). However, it is also generally accepted that strict anhydrous reaction conditions are required for any type of Grignard reactions. However, we found that the reaction of aromatic aldehyde with an allyl halide was highly effective with either THF or water as the reaction solvent but poor in a mixture of THF/water (11) (Table 1). By using aqueous 0.1 N HCl or NH$_4$Cl as the reaction solvent, a quantitative conversion was observed generating a mixture of both the allylation and pinacol coupling products. The use of aqueous 0.1 N NH$_4$Cl as the solvent was found to be superior to the use of 0.1 N HCl and allyl iodide is more effective than allyl bromide. These results suggested that, for organomagnesium reagents, the rate of the two competing reactions involving the hydrolysis of the organometallic reagent and the addition of the organometallic reagent to carbonyl may be comparable, which makes the formation of the carbonyl addition product possible. However, for synthetic purposes the magnesium method will not be as effective as the same methodology using zinc, indium, and tin. Beryllium, the first element in this group, was also examined for the aqueous Barbier-Grignard reaction (12). The results were similar to the magnesium case but showed a lower conversion. Calcium appeared too reactive towards water.

Early transition metals as well as lanthanides were also investigated for the allylation reaction in water. However, only pinacol-coupling products and reduction of the carbonyl were obtained. No allylation product was observed (12). With mid-transition-metals, essentially no reaction was observed under neutral conditions. Under acidic conditions, complicated products including reduction of substrates, pinacol coupling, and allylation were observed (13). Interestingly, a combination of manganese and copper was found to be a highly effective mediator for the allylation of aryl aldehydes in water alone (Eq. 3) (Table 2). No reaction was observed with either manganese or copper alone as the mediator. Only a catalytic amount of copper is required for the reaction. The use of Cu (0), Cu (I), Cu (II) as the copper source are all effective. Under such reaction conditions, better yields of allylation products were obtained with allyl chloride than with allyl bromide or iodide. This was attributed to the formation of Wurtz-type coupling product with the bromide and iodide (14).

Late transition metals and post-transition metals including zinc, tin, indium and bismuth have already been reported as highly effective metals in effecting allylation of carbonyls in water. These reactions generally provide clean allylation products.

While the reason behind such behavior related to different metals in the periodic table is still not clear, it may be due to the small radii of early transition metals which favor inner-sphere single-electron-transfer processes towards

Table 1. Allylation Reaction Mediated by Magnesium in Aqueous Medium

substrate	yield (%)	substrate	yield (%)
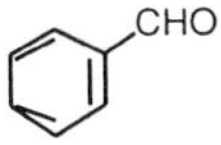 (benzaldehyde) CHO	50	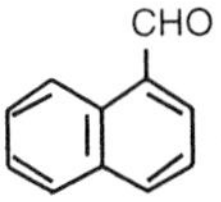 (1-naphthaldehyde) CHO	40
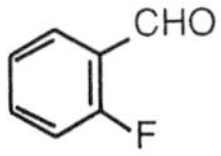 2-F-C6H4-CHO	50	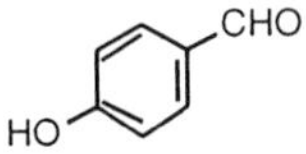 HO- (4-hydroxybenzaldehyde) CHO	21
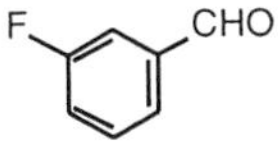 3-F-C6H4-CHO	45	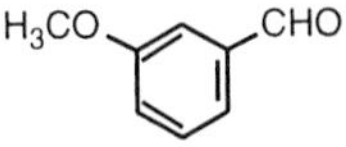 H_3CO- (3-methoxybenzaldehyde) CHO	40
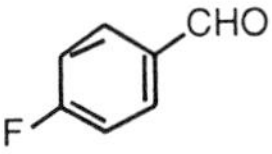 4-F-C6H4-CHO	43	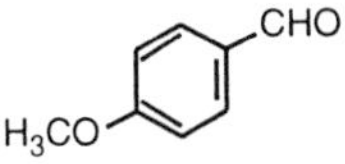 H_3CO- (4-methoxybenzaldehyde) CHO	10
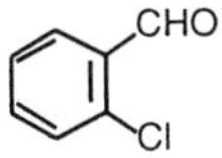 2-Cl-C6H4-CHO	35	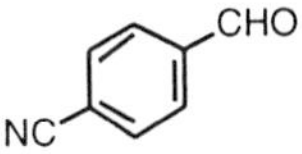 NC- (4-cyanobenzaldehyde) CHO	3
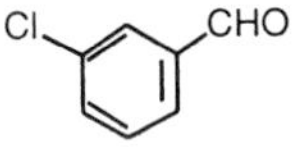 3-Cl-C6H4-CHO	34	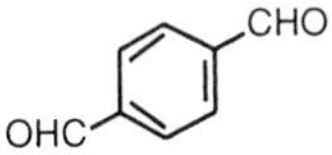 OHC- (terephthalaldehyde) CHO	0
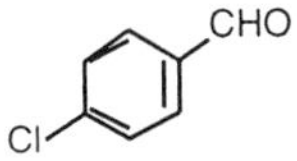 4-Cl-C6H4-CHO	32	 (hexanal) CHO	0
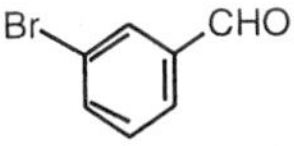 3-Br-C6H4-CHO	30	 (heptanal) CHO	0
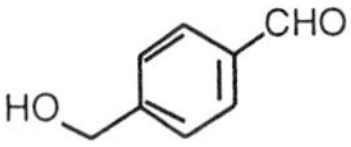 HO- (4-hydroxymethylbenzaldehyde) CHO	45	 (octanal) CHO	0

Yields are referred to isolated ones. Reactions were carried out by using aldehyde: allyl iodide: magnesium turning(1: 3: 10) in 0.1 N aqueous NH_4Cl.

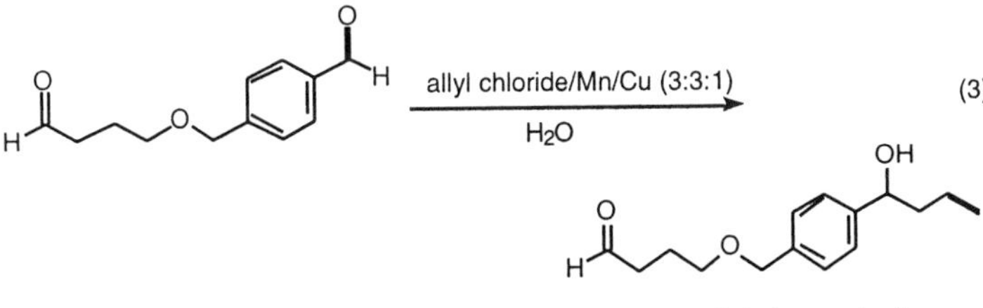

allyl chloride/Mn/Cu (3:3:1)
H_2O

(3)

(only allylation product)

Table 2. Allylation of Aldehydes Mediated by Mn-Cu

substrates	yield(%)[a]	substrates	yield(%)[a]
	83		59
	72		54[b]
	74		0
	62		42
	81[b]		78
	68[b]		74
	51[b]		65
	66[b]		63[b]

Reaction conditions: allyl chloride/Mn/Cu (3 : 3 : 0.1), stirring overnight at room temperature in water. a: isolated yield; b: in H_2O-THF (4:1).

carbonyls while the large radii of late and post-transition metals which favor outer-sphere single-electron-transfer processes towards organic halides (Fig. 1). Thus it appears the key for the success of such reactions in water is to enable the carbon-halogen bond react first.

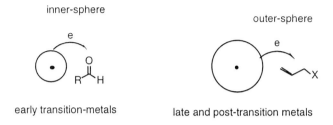

inner-sphere

outer-sphere

early transition-metals   late and post-transition metals

Figure 1.

Metal-Mediated Carbonyl Allylation in Liquid Carbon Dioxide

The reaction between an organometallic reagent and carbon dioxide is a standard method for synthesizing carboxylic acids (10). However, with the success of related reactions in water, we also explored the feasibility of conducting metal-mediated allylation of aldehydes using liquid carbon dioxide as the solvent. We initially had considered attempting the allylation in supercritical CO_2. Despite an established history as an effective extractor, in recent years, much attention has been paid to the use of supercritical CO_2 as a reaction medium (15). It has been proven to be an effective solvent for a number of reactions (16), including the Diels-Alder reaction (17), the Pauson-Khand reaction (18), and free radical brominations (19). Because the temperature and pressure required for a liquid phase CO_2 system are milder, and setting up such a system in the laboratory was relatively easy, we examined allylation of aldehydes in liquid CO_2. Based upon the success of indium-mediated reactions in water, we studied indium metal as a mediator in coupling reactions between carbonyl compounds and allyl bromide in liquid carbon dioxide (Eq. 4). Various studies in the literature (20) attest to the ability of liquid CO_2 to substitute effectively for supercritical CO_2 in different chemical processes.

$$\underset{R}{\overset{O}{\|}}\underset{H}{} + \diagdown\diagup\!X \xrightarrow[\text{liq. } CO_2]{M} \underset{R}{\overset{OH}{\|}}\diagup\diagdown \qquad (4)$$

Table 3. Indium-Mediated Allylation of Aldehydes in Liquid CO_2

Aldehyde	Product Yield(%)	Aldehyde	Product Yield(%)
	82		55
	45		38
	74		69
	58		58
	57		59

Reactions were carried out at room temperature for 48h with the ratio of reactants as: carbonyl/halide/indium=1/1.5/1.5.

The problem with which we were initially confronted was one of system design. The exigencies of working with a high-pressure system imposed certain constraints on the reaction; instead of adding our reactants to the solvent, as is normal for the organic chemist, it was necessary for us to reverse this order. To begin the reaction, we mixed together in the metal bomb the carbonyl compound, indium, and allyl bromide, and only then added liquid CO_2 to the system. There was some concern initially that the allylation reaction would begin before we had even added the liquid CO_2. However, it had also been established in that study that vigorous stirring was necessary to promote the neat allylation (see the next section), and this presented us with a window of opportunity. While the bomb was filling with liquid CO_2, the entire vessel was completely submerged in an ice-water bath. Hopefully, the lower temperature would help impede any neat reaction of the mixture that might possibly occur. Stirring was started only once the vessel was filled with liquid CO_2.

Various carbonyl compounds were examined in which allylation products were obtained smoothly (Table 3). Some of the aldehydes reacted more quickly than others; to ensure completion in each case, the mixture was generally stirred a total of 48 hours. Aliphatic and aromatic aldehydes gave similar yields. For the most part, the presence of substituents on the benzene ring did not have an appreciable effect on the yield or rate of the reaction. In nearly all cases, conversions greater than 85% were observed. We should point it out that all reactions were clean and no appreciable by-product was detected in each case. Some yields were lowered mostly due to product isolation related to ventilation of the apparatus. In the case of a solid substrate such as 4-cyanobenzaldehyde, the use of a surfactant is necessary to effect the allylation.

Metal-Mediated Carbonyl Additions under Neat Conditions

Recently, there has been considerable interest in the study of solventless conditions for organic reactions and synthesis (21). However, under solvent free conditions, all reactants are in close contact and formation of various by-products might be expected in the Barbier-Grignard processes due to the complex nature of such a reaction process. Through our study of metal-mediated reactions under solvent-free conditions, we found that the indium-mediated coupling between carbonyl compounds and allyl bromides under solventless conditions is very facile, giving the cross-coupled product selectively (Eq. 5). Both tin and zinc were found to also mediate this reaction. However, longer reaction times were generally required, and the yields obtained were not as high as (the case) with indium. Also exothermic polymerization seemed to be facilitated in a number of the reactions. This occurred despite lowering the temperature of the reaction in an ice-water bath. Though the yields in the tin reactions were not optimized, the reactivity of the tin seems to parallel that of zinc.

The allylation with allyl bromide was then carried out on a variety of carbonyl compounds (Table 4). In all cases during our investigation, the corresponding products were obtained smoothly with very little by-product being detected, except for furfural, which polymerized rapidly. Aliphatic aldehydes gave yields comparable to those of aromatic aldehydes (Entries 6 and 13), and cyclohexanone reacted similarly (Entry 9). The reaction of an α,β-unsaturated aldehyde gave the 1,2-addition product exclusively (Entry 13). For the aromatic aldehydes, the presence of various substituents, e.g., a methyl, a methoxy, or a bromo group, on the aromatic ring did not have a significant effect on the rate or yield of the reaction.

$$R\overset{O}{\underset{H}{\parallel}} + \overset{}{\diagdown}X \xrightarrow[\text{neat}]{M} R\overset{OH}{\diagup}\diagdown \qquad (5)$$

Table 4. Metal-Mediated Allylation of Carbonyl Compounds under Neat Conditions

Entry	Carbonyl Compound	Metal	Product	Yield(%)
1		In		88
2		Zn		67
3		Sn		60
4		In		79
5		Zn		36
6		In		62
7		In		72
8		In		78
9		In		72
10		In		83
11		In	polymerized	
12		In		60
13		In		66

Reactions were carried out at room temperature for 1h with the ratio of reactants as: carbonyl compound/halide/indium=1/2/1.5.

Mechanism

The allylation of carbonyl compounds with an organometallic reagent is generally regarded as proceeding through the following mechanism (Scheme 1). Although a six-membered ring transition-state has been accepted as a standard in the carbon-carbon bond formation stage (22), the way the allylmetal species is formed is still the subject of controversy. It is generally assumed that the reaction begins with a single-electron-transfer (SET) mechanism involving a radical or a radical anion (23).

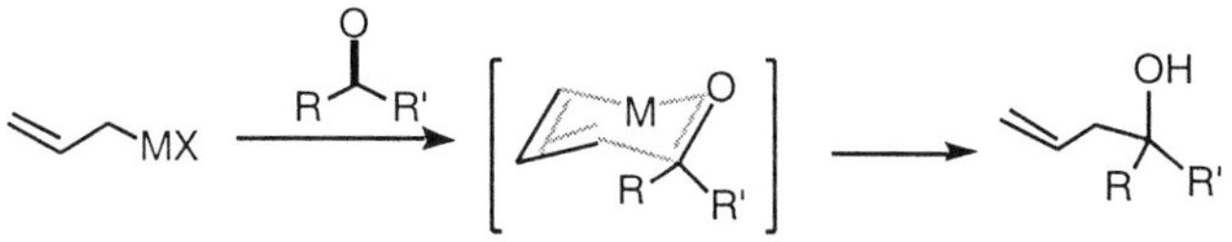

Scheme 1.

If indeed that is the case, then organic halides with lower reduction potentials than allyl halides such as benzylic halides and α-halo carbonyl compounds should be equal if not more reactive under similar reactions conditions. Thus, cross-coupling was examined between an aldehyde and benzylic bromide, as well as between an aldehyde and α-halo carbonyl compounds. However, no reaction was observed in either case under a variety of conditions including higher reaction temperatures (Scheme 2).

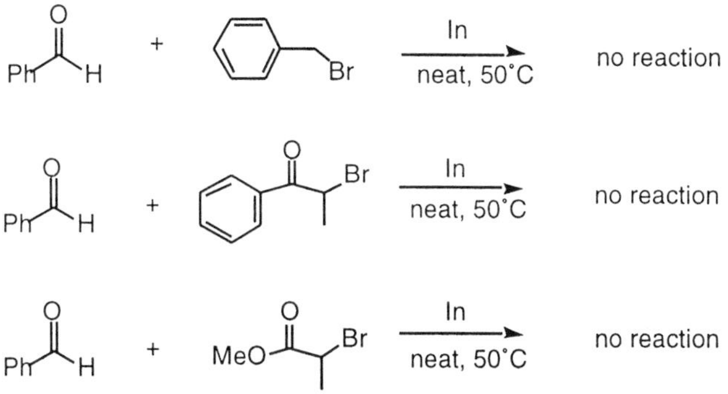

Scheme 2.

These interesting observations led us to consider alternative pathways for the formation of allylmetal species from the allyl halide and metal. We thus tentatively propose a concerted process for the formation of allylmetal species (Scheme 3). It is also possible to have a concerted mechanism involving two metal-atoms. In the latter case, an organometal (I) species will be formed. Interestingly, Chan and co-worker observed the formation of an allylindium (I) intermediate by ^{1}H NMR when studying indium mediated allylation of aldehydes in water (24). With the concerted mechanism, propargyl halide would also be reactive, for example, in water because the structural similarity between allyl halides and propargyl halides. In this proposed mechanism, it would be expected that no reaction would occur when the allyl bromide is absent. This is indeed the case under neutral conditions and without any promoter. Such a modified mechanism can also explain the lack of reactivity with benzyl bromide and α-halo carbonyl compounds. In the former case, a concerted process would break the aromaticity, whereas in the latter cases, a higher activation-energy is required for indium to attack the carbonyl oxygen. However, the existence of alternative mechanisms (e.g., SET or nucleophilic substitution) to generate the allylmetal intermediates is also possible with different metals and under different reaction conditions.

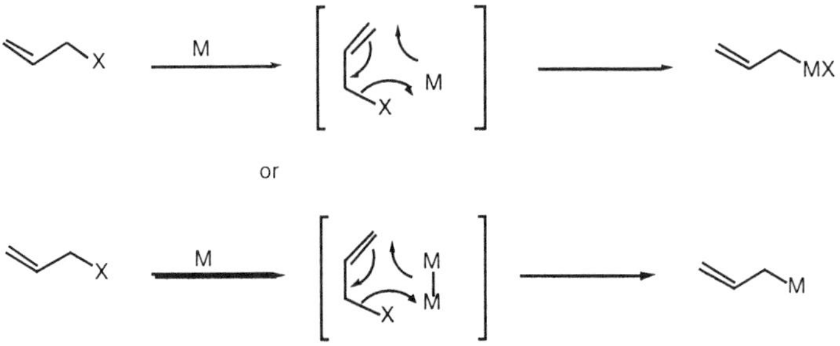

Scheme 3. Proposed Mechanism for Formation of Allylmetal Intermediates

In summary, metal-mediated allylation of carbonyl compounds using water and liquid carbon dioxide as solvents, as well as under neat conditions are shown to be quite effective in most cases. They present relatively clean and efficient methods of forming homoallyl alcohols. A possible concerted process for the formation of allylmetal species from allyl halide and metal may be accountable for the higher reactivity of allyl halides compared to other halides.

Acknowledgments: We also thank the support by Tulane University, by the American Chemical Society (the Petroleum Research Fund), the Louisiana

Board of Regents, the NSF Career Award, and the NSF/EPA Technology for a Sustainable Environment Program.

References

1. *Green Chemistry, Frontiers in Benign Chemical Syntheses and Processes*, Eds. Anastas, P. T.; Williamson, T. C. Oxford University Press, New York, 1998.
2. Frankland, E. *Ann.* **1849**, *71*, 171; Reformatsky, A. *Ber.* **1887**, *20*, 1210.
3. Barbier, P. *Compt. Rend.* **1899**, *128*, 110; Grignard, V. *Compt. Rend.* **1900**, *130*, 1322.
4. Wittig, G. *Newer Methods of Preparative Organic Chemistry* **1948**, 571.
5. Nokami, J.; Otera, J.; Sudo, T.; Okawara, R. *Organometallics* **1983**, *2*, 191.
6. Petrier, C.; Luche, J. L. *J. Org. Chem.* **1985**, *50*, 910.
7. Li, C. J.; Chan, T. H. *Tetrahedron Lett.* **1991**, *32*, 7017; Kim, E.; Gordon, D. M.; Schmid, W.; Whitesides, G. M. *J. Org. Chem.* **1993**, *58*, 5500; Paquette, L. A. in *Green Chemistry, Frontiers in Benign Chemical Syntheses and Processes*, Eds. Anastas, P. and Williamson, T. C. Oxford University Press, New York 1998; Prenner, R. H.; Binder, W. H.; Schmid, W. *Libigs Ann. Chem.* **1994**, 73; Li, X. R.; Loh, T. P. *Tetrahedron: Asymmetry* **1996**, *7*, 1535.
8. Wada, M.; Ohki, H.; Akiba, K. Y. *J. Chem. Soc., Chem. Commun.* **1987**, 708; Katritzky, A. R.; Shobana, N.; Harris, P. A. *Organometallics* **1992**, *11*, 1381.
9. Li, C. J. *Tetrahedron* **1996**, *52*, 5643; Li, C. J.; Chan, T. H. *Tetrahedron* **1999**, *55*, 11149.
10. Wakefield, B. J. *Organomagnesium Methods in Organic Chemistry*, Academic Press, 1995.
11. Li, C. J.; Zhang, W. C. *J. Am. Chem. Soc.* **1998**, *120*, 9102.
12. Keh, C. C. K.; Li, C. J. unpublished results.
13. Li, C. J., *PhD Dissertation*, McGill University 1992.
14. Lexa, D.; Saveant, J. M.; Schaefer, H. J.; Su, K. B.; Vering, B.; Wang, D. L. *J. Am. Chem. Soc.* **1990**, *112*, 6162.
15. For recent reviews, see: Kendall, J. L.; Canelas, D. A.; DeSimone, J. M. *Chem. Rev.* **1999**, *99*, 543; Jessop, P. G.; Ikariya, T.; Noyori, R. *Chem. Rev.* **1999**, *99*, 475; Brennecke, J. F.; Chateauneuf, J. E. *Chem. Rev.* **1999**, *99*, 433; Buelow, S.; Dell'Orco, P.; Morita, D.; Pesiri, D.; Birnbaum, E.; Borkowsky, S.; Brown, G.; Feng, S.; Luan, L. Morgenstern, D.; Tumas, W. in *Green Chemistry, Frontiers in Benign Chemical Syntheses and Processes*, Eds. Anastas, P. T.; Williamson, T. C. Oxford University Press, New York, 1998, p. 265.

190

16 . Kaupp, G. *Angew. Chem. Int. Ed. Engl.* **1994**, *33*, 1452; Wai, C. M.; Hunt, F.; Ji, M.; Chen, X. *J. Chem. Ed.* **1998**, *75*, 1641.
17. Hyatt, J. A. *J. Org. Chem.* **1984**, *49*, 5097.
18. Jeong, N.; Hwang, S. H.; Lee, Y. W.; Lim, J. S. *J. Am. Chem. Soc.* **1997**, *119*, 10549.
19 . Tanko, J. M.; Blackert, J. F. *Science* **1994**, *263*, 203.
20. Hawthorne, S. B. *Anal. Chem.* **1990**, *62*, 633A; Hyatt, J. A. *J. Org. Chem.* **1984**, *49*, 5097; Atwani, Z. *Angew. Chem. Int. Ed. Engl.* **1980**, *19*, 623; Laintz, K. E.; Hale, C. D.; Stark, P.; Rouquette, C. L.; Wilkinson, J. *Anal. Chem.* **1998**, *70*, 400; Isaacs, N. S.; Keating, N. *J. Chem. Soc., Chem. Commun.* **1992**, 876.
21. For a recent review on solvent-free reactions, see: Dittmer, D. C. *Chem. Ind.* **1997**, 779; Loupy, A.; Petit, A.; Hamelin, J.; Texier-Boullet, F.; Jacquault, P.; Mathe, D. *Synthesis* **1998**, 1213; Deshayes, S.; Liagre, M.; Loupy, A.; Luche, J.-L.; Petit, A. *Tetrahedron* **1999**, *27*, 10851; Varma, R. S. *Green Chemistry* **1999**, *1*, 43; Varma, R. S. in *ACS Symp. Ser. 767, Green Chemical Syntheses and Processes*, 292, (2000); Tanaka, K.; Toda, F. *Chem. Rev.* **2000**, *100*, 1025; Metzger, J. O. *Angew. Chem. Int. Ed. Engl.* **1998**, *37*, 2975.
22. Far a recent review about allylation reaction of carbonyl compounds with allylmetal reagents, see: Yamamoto, Y.; Asao, N. *Chem. Rev.* **1993**, *93*, 2207.
23. Molle, G.; Bauer, P. *J. Am. Chem. Soc.* **1982**, *104*, 3481.
24. Chan, T. H.; Yang, Y. *J. Am. Chem. Soc.* **1999**, *121*, 3228.

Chapter 14

Micellar Catalysis as a Clean Alternative for Selective Epoxidation Reactions

J. H. M. Heijnen, V. G. de Bruijn, L. J. P. van den Broeke, and J. T. F. Keurentjes

Department of Chemical Engineering and Chemistry, Process Development Group, Eindhoven University of Technology, P.O. Box 513, 5600 MB Eindhoven, The Netherlands

In the search for more environmentally friendly processes, the use of micellar structures opens up a range of new possibilities. One of the most exciting applications is the incorporation of homogeneous catalysts in micelles to perform reactions: micellar catalysis. Micellar catalysis can be used to perform for example epoxidation reactions. In this way organic solvents, in which epoxidations are usually performed, are replaced by aqueous surfactant solutions. In this study propylene and 1-octene have been successfully epoxidized by hydrogen peroxide, catalyzed by micelle-incorporated porphyrin catalysts. Furthermore, several experimental techniques have been used to get more insight in the micelle-catalyst system.

Introduction

In micellar catalysis, a nonpolar homogeneous catalyst is solubilized in an aqueous micellar system, where it converts nonpolar reactants to polar products. To study micellar catalysis, the epoxidation of propylene to propylene oxide has been chosen as a model reaction. Hydrogen peroxide is used as the oxidizing agent:

$$CH_3-CH=CH_2 \;+\; H_2O_2 \;\longrightarrow\; CH_3-\overset{\displaystyle O}{\overset{\diagup\diagdown}{CH-CH_2}} \;+\; H_2O$$

The principle of micellar catalysis of propylene to propylene oxide is schematically shown in Figure 1. Propylene gas is bubbled through an aqueous solution, containing the micelle-incorporated epoxidation catalyst and diffuses from the gas phase through the aqueous phase into the micelles. In the micelles, the catalyst converts propylene and hydrogen peroxide to propylene oxide and water. In-situ extraction of polar propylene oxide will reduce its further oxidation, increasing the reaction selectivity. The use of water instead of organic solvents will avoid the emission of the latter. Due to its high heat capacity, water will allow for safer process operation, since boiling conditions are not easily reached. Furthermore, it is known that an organized medium affects reaction rate and selectivity of many organic reactions *(1,2)* as well as metal catalyzed reactions *(3-6)*. Another advantage is that, once incorporated into a micelle, the homogeneous nonpolar catalyst can be easily separated from the polar products, e.g. by micellar enhanced ultrafiltration *(7)*.

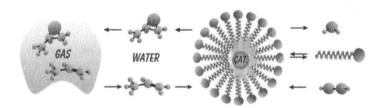

Figure 1. The principle of micellar catalysis.

The goal of this research is to design a process for propylene epoxidation by hydrogen peroxide based on micellar catalysis. Three levels of design can be identified. The first level is the design of the homogeneous epoxidation catalyst. Catalyst performance can be optimized by modification of the ligand and the central transition metal atom *(8-10)*. At the second level, the micelle-catalyst combination is designed for optimal contacting of catalyst and reactants on a molecular scale. Due to the high polarity gradient in the micelle, the nonpolar propylene can be brought into close contact with the polar hydrogen peroxide. By altering substituents on the catalyst ligand, the catalyst polarity can be modified, resulting in a different location in the micelle. Furthermore, the surfactant can be designed by changing the type, length and sequence of the surfactant blocks. Finally, reactant contacting on a macroscale, operating conditions, and product separation are optimized in a process design. Our main focus has been on micelle-catalyst design and process design.

Experimental

The epoxidation catalysts used are shown in Figure 2. Ti(Silsesquioxane) [{(c-C_5H_9)$_7$Si$_7$O$_{12}$}Ti(IV)(η^5-C_5H_5)] was kindly supplied by the Schuit Institute of Catalysis at Eindhoven University of Technology. Mn(Salen)Cl [(+)-N,N,-Bis(3,5-ditert-tubylsalicylidene)-1,2-cyclohexanediamino-manganese(III)-chloride] and Mn(TPP)Cl [5,10,15,20-tetrakis(phenyl)porphyrin Manganese(III) chloride] were purchased from Fluka and Mn(TDCPP)Cl [5,10,15,20-tetrakis(2,6-dichlorophenyl)porphyrin Manganese(III) chloride] was purchased from Frontier Scientific. Propylene (99+%, lecture bottle), acetonitrile, dichloromethane, n-octane and iso-octane were obtained from Aldrich. Hydrogen peroxide (50 wt% in water) and MPEG (polyethylene glycol monomethyl ether) were obtained from Acros Chimica and imidazole and 1-octene were obtained from Merck. The surfactants used are listed in Table I.

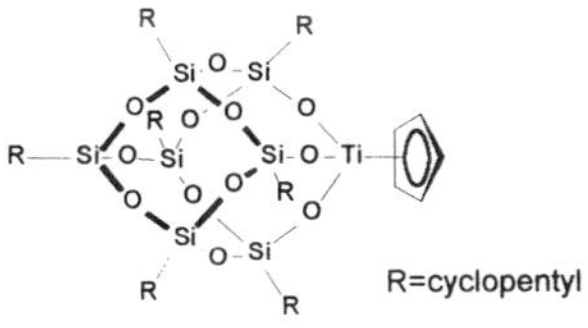

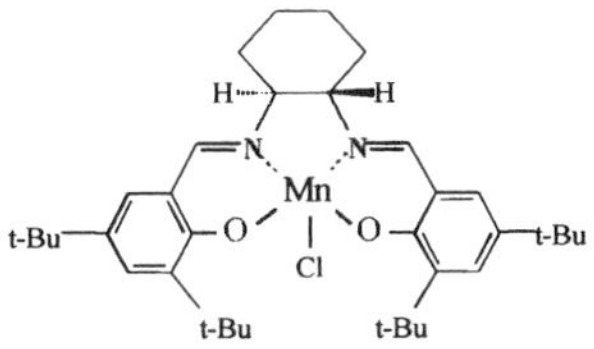

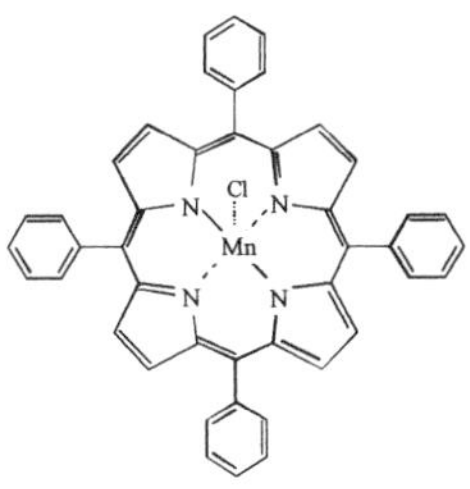

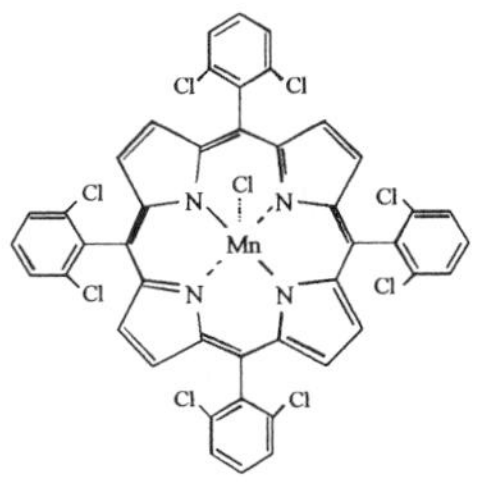

Figure 2. Selected epoxidation catalysts. 1 Ti(Silsesquioxane); 2 Mn(Salen)Cl; 3 Mn(TPP)Cl; 4 Mn(TDCPP)Cl.

Table I. Selected surfactants[a]

Surfactant	Structure	Purchased from
Triton X-100	$C(CH_3)_3-CH_2-C(CH_3)_2-C_6H_4-PEO_{9.5}$	Sigma
Brij35	$CH_3-(CH_2)_{11}-PEO_{23}$	Acros
SDS	$CH_3-(CH_2)_{11}-OSO_3\,Na$	Sigma
CTAB	$CH_3-(CH_2)_{15}-N(CH_3)_3\,Br$	Sigma
Pluronic P84	$PEO_{19}-PPO_{43}-PEO_{19}$	BASF
Pluronic L64	$PEO_{13}-PPO_{30}-PEO_{13}$	BASF
Pluronic P103	$PEO_{17}-PPO_{60}-PEO_{17}$	BASF
Pluronic F127	$PEO_{100}-PPO_{65}-PEO_{100}$	BASF

[a] PEO: polyethylene oxide; PPO: polypropylene oxide.

In the solubilization experiments, the epoxidation catalysts were added to 10 wt% surfactant solutions and stirred for a few hours up to several days. To determine whether solubilization was successful, the solutions were analyzed visually. Successful catalyst solubilization resulted in transparent surfactant solutions, which were often colored due to the presence of the catalyst. Poor or no solubilization at all resulted in dispersions of solid catalyst in the surfactant solution. Without agitation, these solutions showed complete catalyst precipitation. To determine the micelle diameter of the catalyst-surfactant solution, a Coulter N4 plus dynamic light scattering apparatus was used. UV/vis and ^{1}H NMR spectroscopy were performed on a Spectronic Genesys 5 spectrophotometer and a Varian Unity Inova 500 MHz instrument, respectively.

For epoxidation experiments the catalyst was solubilized in a 3.5 wt% Triton X-100 surfactant solution and subsequently imidazole, 1-octene and iso-octane were added. Iso-octane served as an internal standard for GC analysis and imidazole was used as a cocatalyst. The solution was left to equilibrate for a few hours and imidazole coordination was checked by UV/vis spectroscopy. After the hydrogen peroxide solution was injected, the reaction started and liquid phase GC-analysis (Fisons GC 8000 series) was performed every 5 to 10 minutes. For the determination of micelle-water partition equilibria, a glass bulb equipped with a thermometer, septum port, gas inlet and gas outlet was filled with solution. Subsequently, the glass bulb was evacuated by a vacuum pump. Then propylene was added from a lecture bottle. With a pressure regulator, the pressure in the bulb was kept constant. The propylene solubility in both pure water and surfactant solutions was determined by liquid phase GC analysis. Ultrafiltration experiments were conducted in a solvent resistant stirred cell (Millipore) using regenerated cellulose ultrafiltration membranes with a cut-off value of 3 kDa (Millipore).

Results and Discussion

For the catalyst-micelle design, the catalyst solubilization and its influence on the micelle diameter have been investigated. Moreover, the catalyst location has been determined by lattice calculations and by ^1H NMR and UV/vis spectroscopy. For the process design, reaction kinetics and the micelle-water partition equilibria have been determined.

Catalyst-Micelle Design

The primary function of the surfactant is to form aggregates that can incorporate the catalyst. To understand how catalyst reactivity, selectivity and stability depend on the surfactant type, it is important to obtain insight in the catalyst-micelle structure. In that respect, the catalyst location is an important parameter. As an example, two cases are considered in Figure 3.

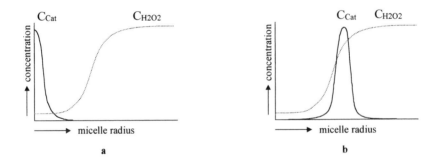

Figure 3. Different catalyst locations in the micelle. **a**: *At the micelle core,* **b**: *At the micelle-water interface.*

Due to the polarity gradient in the micelle, reactant, product and catalyst concentrations change over the micelle radius. If high hydrogen peroxide concentrations maximize reactivity, the optimal catalyst location is at the micelle-water interface (situation **b**). If the catalyst is susceptible to hydrogen peroxide, a catalyst location near the micelle core is preferred to maximize catalyst stability (situation **a**). In the following, catalyst solubilization and catalyst location are assessed both experimentally and computationally.

Catalyst solubilization

The solubilization of the epoxidation catalysts was investigated for eight different surfactant types. The results of the solubilization studies are given in Table II. While the Mn(TPP)Cl and Mn(Salen)Cl catalyst could be readily solubilized, the Ti(Silsesquioxane) catalyst failed to solubilize in any of the surfactant solutions tested. Heating and sonification failed to promote Ti(Silsesquioxane) solubilization. In additional experiments, the Ti(Silsesquioxane) catalyst was dissolved in toluene and successfully emulsified in several surfactant solutions. After stripping of the toluene, however, the catalyst precipitated again. Mn(TDCPP)Cl solubilization depends on the surfactant type. The results seem to indicate that the surfactant should contain a PEO headgroup for successful Mn(TDCPP)Cl solubilization. Clearly catalyst solubilization depends on the interaction between micelle and catalyst, but also the catalyst volume seems to play an important role in the solubilization process.

Table II. Catalyst solubilization

Surfactant solution	1	2	3	4
Triton X-100	-	+	+	+
Brij35	-	+	+	+-
SDS	-	+	+	-
CTAB	-	+	+	-
Pluronic P84[a]	-	+	+	+
Pluronic L64[a]	-	+	+	nt
Pluronic P103[a]	-	+	+	nt
Pluronic F127[a]	-	+	+	nt

1 Ti(Silsesquioxane); 2 Mn(Salen)Cl; 3 Mn(TPP)Cl; 4 Mn(TDCPP)Cl; +: complete solubilization; +-: partial solubilization; -: no solubilization; nt: not tested; [a] experiment conducted at 50°C

The influence of the catalyst concentration on the micelle diameter has been investigated for the Mn(Salen)Cl - Triton X-100 system by dynamic light scattering. In Figure 4 the micelle diameter is given as a function of the catalyst concentration at different temperatures. The micelle diameter decreases with increasing catalyst concentration. This indicates that with an increasing catalyst to surfactant ratio, the catalyst is solubilized in smaller micelles. This process continues until there are not enough surfactant molecules left to stabilize the nonpolar catalyst.

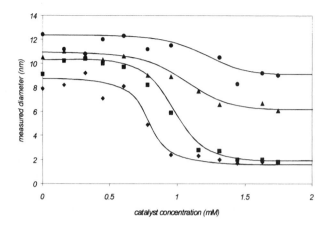

Figure 4. Micelle diameter as a function of the catalyst concentration in 10 wt% Triton X-100. ◆ : 30 °C; ■ : 35 °C; ▲ : 40 °C; ● : 45 °C.

It follows from Figure 4 that the reduction in micelle diameter is less pronounced at higher temperatures. At higher temperatures the PEO chains become more hydrophobic. Thus the water, present in the palisade layer is squeezed out, resulting in a decreased micelle curvature, which induces the formation of larger micelles. At the maximum loading capacity of the micelles, the catalyst precipitates from the solution. This is schematically shown in Figure 5. These measurements indicate that the solubilization process can be regarded as surfactant adsorption onto the catalyst surface.

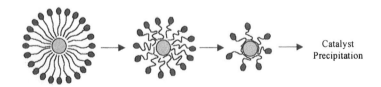

Figure 5. Decreasing micelle diameter at increasing catalyst to surfactant ratio.

Catalyst location

To simulate catalyst solubilization, the lattice-based self-consistent field theory of Scheutjens et al. *(11)* is used. The components in the system are divided into segments of equal volume and their mutual interaction is accounted for by nearest neighbour Flory-Huggins interaction parameters. The components, consisting of single segments or chains of segments, follow step-weighted random walks within a potential field gradient defined by the lattice geometry. The equilibrium segment distribution is found by an iterative process, wherein the total free energy is minimized. Calculations performed for the Mn(Salen)Cl catalyst in spherically symmetric P84 micelles result in radial volume fraction profiles as shown in Figure 6. In the nonionic micelles PEO chains make up the hydrophilic part of the micelle. Since the PEO chains extend into the water phase, the location of the micelle-water interface is somewhat diffuse. From the lattice calculations in Figure 6 it can be seen that beyond the 25th lattice segment no PEO is present. The catalyst is located in the micelle core. The catalyst squeezes some water and PPO out of the core, but the rest of the micelle remains unaffected.

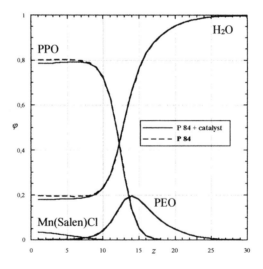

Figure 6. Concentration profiles in volume fractions (φ) over the micelle radius (z) calculated for P84 surfactant solutions with and without Mn(Salen)Cl. The lattice segment length, z, is about 0.2 nm.

UV/vis spectroscopy can be used to determine the catalyst location experimentally *(12)*. The absorption spectrum of the solubilized catalyst is compared to the absorption spectra of the catalyst in solvents that are representative of the micelle core and the micelle shell. This will allow for a rough determination of the catalyst location. A more precise location can be determined when mixtures of solvents are used. The location of Mn(TPP)Cl in Brij35 and Triton X-100 micelles has been determined, using UV/vis spectroscopy. The results are shown in Figure 7. The absorption spectra of the catalyst in the surfactant solutions show a much better similarity with the absorption spectrum in MPEG, as compared to the spectrum in *n*-octane. Addition of water to MPEG results in an even better match. Therefore, it can be concluded that the Mn(TPP)Cl catalyst is located in a MPEG-water environment with a water content higher than 50 vol%. This environment is found in the outer regions of the palisade layer in both Brij35 and Triton X-100 micelles.

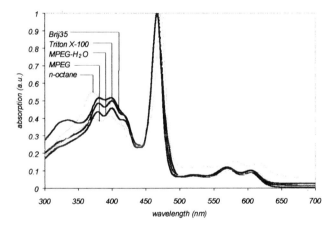

Figure 7. Absorption spectra of Mn(TPP)Cl in 3.5 wt% Brij35, 3.5 wt% Triton X-100, MPEG (methoxypolyethylene glycol, Mw = 350), n-octane, and a 50-50 vol% MPEG-water mixture.

^{1}H NMR can also be used to determine the catalyst location in the micelles *(13-15)*. The spin-lattice relaxation time, T_1, of Pluronic P84 surfactant solutions in D_2O decrease upon solubilization of the Mn(Salen)Cl catalyst. The decrease in T_1 was more pronounced for the nonpolar section of the surfactant as compared to the polar section. This indicates that the Mn(Salen)Cl catalyst is located at the core of the Pluronic P84 micelles, which is consistent with the lattice calculations. Moreover, it was noted that at elevated temperatures, the T_1 of the PPO chains increased and the T_1 of the PEO chains decreased, indicating a shift in the Mn(Salen)Cl location towards the PEO chains. This is consistent with the polarity decrease of PEO chains at increasing temperatures *(16,17)*.

Process Design

The two commercial processes for the epoxidation of propylene suffer from major drawbacks, like corrosive conditions, the use of organic solvents and substantial waste production and co-product formation *(18)*. A process based on micellar catalysis can circumvent these problems. A preliminary process design is shown in Figure 8. Gaseous propylene and liquid hydrogen peroxide are fed into the reactor (a) where they are converted into propylene oxide and water. The gas outlet of the reactor mainly contains propylene and propylene oxide. The liquid outlet mainly consists of propylene oxide, water, hydrogen peroxide, surfactant and catalyst. The catalyst and surfactant are recovered in the ultrafiltration unit (b) by micellar enhanced ultrafiltration. Crude propylene oxide is further purified in distillation columns (c) and excess propylene and hydrogen peroxide are fed back into the reactor. The separation section is rather straightforward since the boiling points of propylene (-48°C), propylene oxide (34°C), water (100°C) and hydrogen peroxide (151°C) are far apart.

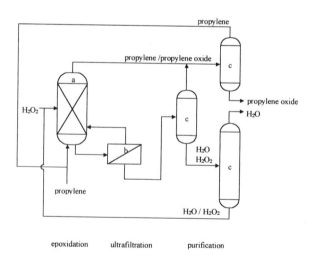

Figure 8. Preliminary process design, a) epoxidation reactor; b) ultrafiltration unit; c) distillation columns.

For the process design several key issues can be identified. First of all, the micelle is an organized medium, which is known to alter reaction rates and selectivities *(1-6)*. Furthermore, the micellar solutions are micro-heterogeneous two-phase media, where the components are distributed between the micelle and the water phase. To determine the intrinsic reaction rate constants in the micelles, partition equilibria must be known. Surfactant accumulation at the gas-liquid interface results in a reduced gas bubble size and since the micelles ar three orders of magnitude smaller than the liquid boundary layer, the micelles can shuttle propylene from the gas-liquid interface to the liquid bulk *(19)*. One of the major drawbacks of homogeneous catalysis is the catalyst recovery. In our concept, the solubilized homogeneous catalyst can be separated from the aqueous phase by micellar enhanced ultrafiltration *(7)*. This is schematically shown in Figure 9. Preliminary ultrafiltration experiments with the Mn(TPP)Cl catalyst solubilized in Triton X-100 solution show that the catalyst can be retained with high selectivity.

Micelle-water partition equilibria

Micelle-water partition equilibria have been determined for propylene in several surfactant solutions. From the saturation concentrations in pure water and the surfactant solution combined with the micelle hold-up, the partition equilibrium can be calculated. The micelle-water partition equilibria are depicted as a function of temperature in Figure 10.

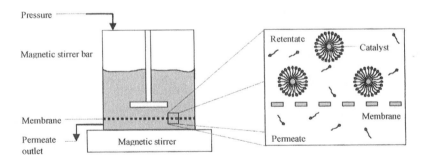

Figure 9. Micellar enhanced ultrafiltration. The micelle-incorporated catalyst is retained by the ultrafiltration membrane.

Due to the polarity change of the PEO-groups, the micelle-water partition equilibria for propylene increase with temperature. Moreover, Brij35, having the most nonpolar tail, has the highest affinity for propylene. Future research will be focused on the determination of micelle-water partition equilibria of hydrogen peroxide and propylene oxide.

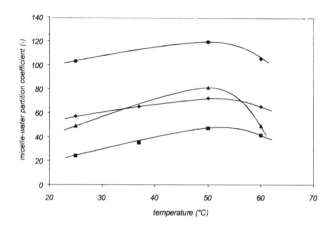

Figure 10. Propylene micelle-water partition coefficients as a function of temperature. ■: *Pluronic P84;* ▲: *Pluronic F127;* ◆: *Triton X-100;* ●: *Brij35.*

Reaction kinetics in micelles

Screening experiments with the Mn(Salen)Cl, Mn(TPP)Cl and Mn(TDCPP)Cl solubilized in micellar Triton X-100 solutions show that the porphyrins exhibit the highest catalytic performance. Therefore, kinetic experiments have been conducted with the porphyrin catalysts. As 1-octene and propylene epoxidations by porphyrins show similar reactivities, 1-octene was used for catalyst screening because of its ease of handling.

The reaction mechanism of alkene epoxidation by porphyrin catalysts using hydrogen peroxide as oxidizing agent and imidazole as cocatalyst is given below *(8)*. Epoxidation reactions catalyzed by porphyrins using hydrogen peroxide require a cocatalyst for catalyst activation. Often imidazole is used as a cocatalyst. The porphyrin is activated by hydrogen peroxide to yield the oxo-species and subsequently epoxidizes the alkene. Imidazole serves both as a base and axial ligand.

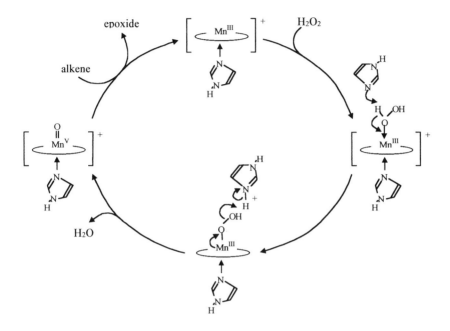

With the Mn(TDCPP)Cl catalyst, 1-octene is epoxidized by hydrogen peroxide. The 1-octene oxide yield is shown in Figure 11 as a function of time. The initial turnover frequency is 250 h^{-1}. After 30 minutes the reaction stops, but after the addition of 5.5 μmole imidazole the reaction proceeds again. The same results have been obtained when sodium hydroxide was added instead of imidazole. Moreover, during the reaction the pH decreased. A pH decrease results in imidazole protonation. The positively charged protonated imidazole will fail to coordinate to the porphyrin catalyst and thus the reaction comes to a stop. We have strong indications the hydroxyl group at the end of the PEO headgroup of Triton X-100 is oxidized to a carboxylic group, resulting in the pH decrease. The pH decrease can also be the result of the oxidation of the alkaline imidazole, which has been reported for organic solvents (9,20). Since imidazole and alkene compete for the active oxo-species, undesired imidazole oxidation can be reduced by higher alkene concentrations. In Table III, an overview is given of 1-octene epoxidations carried out with the two solubilized porphyrin catalysts.

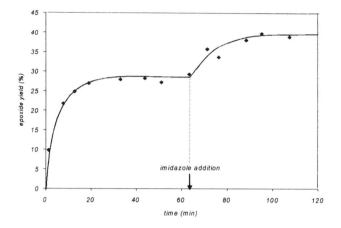

Figure 11. 1-Octene epoxidation by Mn(TDCPP)Cl in 3.5 wt% Triton X-100;
$C_{Mn(TDCPP)Cl} = 0.24$ mM, $C_{imidazole} = 5.6$ mM, $C_{1-octene} = 9.2$ mM, $C_{H2O2} = 96$ mM.

Table III. Epoxidation of 1-octene with H_2O_2 in micellar systems

#	Catalyst	Medium	Cocatalyst	TOF $[h^{-1}]^a$	Stability
1	Mn(TPP)Cl	Brij35 [b]	-	10.5	-
2	Mn(TPP)Cl	Brij35 [b]	imidazole	30	-
3	Mn(TPP)Cl	SDS [b]	imidazole	30	-+
4	Mn(TPP)Cl	Triton X-100 [b]	imidazole	82.5	-
5	Mn(TPP)Cl	DCM/CH₃CN [c]	imidazole	1.1	--
6	Mn(TDCPP)Cl	Triton X-100 [b]	-	10	~ 100 %
7	Mn(TDCPP)Cl	Triton X-100 [b]	imidazole	250	~ 100 %
8	Mn(TDCPP)Cl	DCM/CH₃CN [c]	imidazole	392	~ 100 %
9	-	Brij35 [b]	imidazole	0	
10	-	Triton X-100 [b]	imidazole	0	

[a] Initial turnover frequency; [b] 3.5 wt% surfactant solutions; [c] 50-50 vol%.

It can be concluded that Mn(TPP)Cl is less active and less stable than Mn(TDCPP)Cl, which is in agreement with literature *(8,21)*. Mn(TDCPP)Cl gives excellent results, apart from the pH decrease. Detailed experiments show that the epoxidation catalyzed by Mn(TPP)Cl in micellar media is first order in hydrogen peroxide and zero order in 1-octene. This indicates that the formation of the active oxo-species is the rate-limiting step. This is in agreement with alkene epoxidations catalyzed by porphyrins in organic solvents *(21)*. The Mn(TPP)Cl is more stable and more reactive in the micellar solutions as compared to the organic solvent. This is in agreement with results obtained by Monti et al. *(6)*. Mn(TDCPP)Cl, however, is less reactive in the micellar solution than in the organic medium. The differences in reactivity and stability between the micellar solutions and the organic solvent can be explained as follows. Hydrogen peroxide has a higher activity in organic media, as compared to micellar solutions, due to the absence of stabilizing water molecules *(21)*. Moreover, the organic solvent is a one-phase medium. Micellar solutions, however, are two-phase media, where the species are distributed between the micelle phase and the water phase. Since both hydrogen peroxide and imidazole are hydrophilic, they prefer the water phase over the micelle phase. This results in decreased hydrogen peroxide and imidazole concentrations in the micelle, at the locus of reaction, thus retarding the 1-octene epoxidation.

It is noted that the surfactant structure has a substantial influence on the performance of the Mn(TPP)Cl catalyst. The Triton X-100 solution shows increased activity, while the SDS solution shows improved catalyst stability. These results imply that the choice of an optimal surfactant type can strongly enhance catalyst performance.

Concluding Remarks

Propylene and 1-octene have successfully been epoxidized by hydrogen peroxide, catalyzed by a micelle-incorporated porphyrin catalyst. Several experimental techniques have been used to get more insight in the micelle-catalyst system. The surfactant type has a considerable effect on catalyst solubilization as well as on catalyst performance. Furthermore, a decrease in the micelle diameter has been observed upon catalyst solubilization. The catalyst location has been determined in several catalyst-surfactant systems, using UV/vis and NMR measurements and lattice calculations. The micelle-incorporated catalyst could be retained with high retention by an ultrafiltration membrane. Based on the foregoing, a preliminary process design has been presented for propylene epoxidation by hydrogen peroxide, catalyzed by a micelle-incorporated porphyrin catalyst. The research on catalyst location, catalyst solubilization, micellar enhanced ultrafiltration and reaction kinetics will be extended to new catalyst-surfactant combinations. Detailed reaction kinetics will be determined and more robust co-catalysts and surfactants will be tested. If the mechanisms affecting the performance of the system are understood, the system will be optimized by tuning the catalyst, the surfactant

206

and the process. Finally, it is anticipated that the application of micellar catalysis is not just limited to epoxidations, but can be applied to a much broader range of reactions.

Acknowlegdements

We thank Eka Chemicals and ABB Lummus for financial support and F.A.M. Leermakers of Wageningen University for his help with the lattice calculations.

References

1. Fendler, J.H. *Membrane Mimetic Chemistry;* Wiley: New York, NY, 1982.
2. Menger, F.M.; Portnoy, C.E. *J. Am. Chem. Soc.* **1967,** *89,* 4698-4703.
3. Rabion, A.; Buchanan, R.M.; Seris, J.; Fish, R. *J. Mol. Cat. A: Chem.* **1997,** *116,* 43-47.
4. Karakanov, E.A.; Flippova, T.Yu.; Martynova, S.A.; Maximov, A.L.; Predeina, V.V.; Topchieva, I.N. *Catalysis Today* **1998,** *44,* 189-198.
5. Broxton, T.J.; Nasser, A. *Can. J. Chem.* **1997,** *75,* 202-206.
6. Monti, D.; Tagliatesta, P.; Mancini, G.; Boschi, T. *Angew. Chem. Int. Ed.* **1998,** *37,* 1131-1133.
7. *Surfactant-Based Separation Processes;* Scamehorn, J.F.; Harwell, J.H., Eds.; Surfactant Sci. Series; Marcel Dekker: New York, NY, 1989; Vol. 33.
8. Battioni, P.; Renaud, J.P.; Bartoli, J.F.; Reina-Artiles, M.; Fort, M.; Mansuy, D. *J. Am. Chem. Soc.* **1988,** *110,* 8462-8470.
9. Anelli, P.L.; Banfi, S.; Legramandi, F.; Montanari, F.; Pozzi, G.; Quici, S.; *J. Chem. Soc. Perkin Trans.* **1993,** *1,* 1345-1357.
10. Banfi, S.; Cavalieri, C.; Cavazzini, M.; Trebicka, A. *J. Mol. Cat. A: Chem.* **2000,** *151,* 17-28.
11. Scheutjens, J.M.H.M.; Fleer, G.J. *J. Phys. Chem.* **1979,** *83,* 1619-1635.
12. Schenning, A.P.H.J.; Hubert, D.H.W.; Feiters, M.C.; Nolte, R.J.M. *Langmuir* **1996,** *12,* 1572-1577.
13. Madzumdar, S. *J. Phys. Chem.* **1990,** *94,* 5947-5953.
14. Gandini, S.C.M.; Yushmanov, V.E.; Borissevitch, I.E.; Tabak, M. *Langmuir* **1999,** *15,* 6233-6243.
15. Maiti, N.C.; Madzumdar, S.; Periasamy, N. *J. Phys. Chem.* **1995,** *99,* 10708-10715.
16. Karlström, G.J. *J. Phys. Chem.* **1985,** *89,* 4962-4964.
17. Björling, M.; Linse, P.; Karlström, G. *J. Phys. Chem.* **1990,** *94,* 471-481.

18. *Kirk-Othmer Encyclopedia of Chemical Technology, 4*th *edition;* Wiley: New York; Vol. 20, pp 271-302.
19. Mehra, A. *Current Science* **1990,** *59,* 970-979.
20. Legemaat, G. Ph.D. thesis, Utrecht University, Utrecht, Netherlands, 1990.
21. Thellend, A.; Battioni, P.; Mansuy, D. *J. Chem. Soc., Chem. Commun.* **1994,** 1035-1036.

Chapter 15

Aqueous Polyglycol Solutions as Alternative Solvents

Neil F. Leininger, Reid Clontz, John L. Gainer[*], and Donald J. Kirwan

Department of Chemical Engineering, University of Virginia,
Charlottesville, VA 22904–4741

In this work, polyethylene glycol 300 (PEG 300), polypropylene glycol 425 (PPG 425), and their aqueous solutions have been studied as possible alternative solvents for fine chemical synthesis. These solvents are non-volatile, eliminating the possibility of fugitive emissions, and are relatively non-toxic. Properties of the polyglycol-H_2O solutions were studied. Two enzymatic organic reactions were studied in the alternative solvents, an enzymatic esterification and an enzymatic hydrolysis reaction. From these reactions, it was determined that water is not bound strongly to the polyglycols and is free to react. Four non-catalytic, homogeneous reactions: a S_N1 reaction, two Diels-Alder reactions, and a S_N2 reaction were also studied. Rate constants were obtained and compared to those obtained in traditional organic solvents. Favorable rate constants were obtained for the S_N1 and Diels-Alder reactions. In addition, many organic compounds are quite soluble in these alternative solvents, so significant overall reaction rates can be achieved.

In recent years, the demand for safe, environmentally-friendly solvents has greatly increased. This has been spurred by new EPA regulations. Regulations such as the Montreal Protocol, the Superfund Amendments and Reauthorization Act (SARA) of 1986, the 1990 Clean Air Act Amendments, and others have forced chemical companies to reduce waste emissions to the environment (1,2). Other safety regulations, such as those established by the Occupational Safety and Health Administration (OSHA) to protect workers from exposure to harmful solvents, have also increased the demand for alternative solvents (1). In addition, the need for safe, alternative solvents has economic ramifications. In order to meet the regulations imposed by the EPA and OSHA, a significant amount of money is spent every year. It is estimated that as much as $115 billion dollars is spent on pollution control and treatment every year (3).

Despite the money and effort spent on pollution control and treatment in 1997, over 2.5 billion pounds of waste were still released to the environment (4). There were significant waste releases in the pharmaceutical industry alone, with 29.9 million pounds released to the environment in 1995 (5). This clearly shows the need for the development of less-volatile, environmentally benign solvents.

We are investigating situations where polyglycol solutions can be used as alternative solvents. In particular, polyethylene glycol (PEG) 300 and polypropylene glycol (PPG) 425 are attractive. First of all, both PEG 300 and PPG 425 are non-volatile. This is important because of the 2.5 billion pounds of chemical waste released to the environment in 1997, air emissions accounted for 52 % of the emissions (4). In the pharmaceutical industry, of the 29.9 million pounds of waste released to the environment in 1995, air emissions accounted for 17.2 million pounds or 57 % of the emissions (5). Interestingly, two of the more common solvents used in the pharmaceutical industry, methanol and dichloromethane, accounted for 11.6 million pounds of the air emissions (5). Therefore, the use of non-volatile solvents should drastically cut down on the amount of air emissions in the pharmaceutical industry and the chemical industry in general. From a health perspective, a non-volatile solvent greatly reduces the chances of fire or explosion and decreases worker exposure to solvent vapors.

Secondly, PEG 300 and PPG 425 are both relatively non-toxic. PEG 300 is currently used in various biological products and medicines (6), and is certified as safe for ingestion (6, 7). PPG 425 is more toxic (8), but appears to be much safer than commonly-used solvents.

In this study, PEG 300 and PPG 425 solutions were investigated and possible uses as well as limitations were identified. In an effort to do this, physical properties of these solutions were investigated and organic compound solubilities were determined. Also, enzymatic reactions and homogeneous, non-catalytic reactions were conducted using the polyglycol solutions. For the non-catalytic reactions, rate constants were obtained and compared to those in traditional organic solvents.

Molecular Interactions

Molecular interaction parameters were obtained for the polyglycol-H_2O solutions. Polarity, hydrogen bond-accepting, and hydrogen bond-donating measurements were made following a protocol developed previously by Kamlet and co-workers (9,10,11,12). All of these measurements were preformed using solvatochromic dyes. To obtain the polarity data, solvatochromic dye 4-nitroanisole was used. To obtain hydrogen bond-accepting parameters, solvatochromic compounds 4-nitroanisole and 4-nitrophenol were used. For the hydrogen bond-donating parameters, 4-nitroanisole and Reichardt's dye, (2,6-diphenyl-4-(2,4,6-triphenylpyridinio)-phenolate) were used. All measurements were performed using either a Hewlett Packard, Model 8452A diode array spectrophotometer or a Beckman, DU-7 spectrophotometer. As a check, molecular interaction values for some common solvents were also obtained, and it was found that our values were within 10% of the published values of Kamlet and co-workers (9,10,11,12).

The molecular interaction parameters for cyclohexane (9), methanol (9), PEG 300, and PPG 425 are shown in Table 1. Table 1 shows that both PEG 300 and PPG 425 are reasonably polar molecules, as both solvents have polarity values greater than that of methanol, with PEG 300 being more polar than PPG 425. Both PEG 300 and PPG 425 have hydrogen bond-accepting capabilities that are similar to methanol, but their hydrogen bond-donating abilities are less than that of methanol.

Table I. Molecular Interaction Parameters

Solvent	Polarity (π^*)	HBA (β)	HBD (α)
Cyclohexane (9)	0.00	0.00	0.00
Methanol (9)	0.60	0.62	0.93
PEG 300	0.94	0.60	0.45
PPG 425	0.65	0.67	0.34

NOTE: HBA = Hydrogen bond accepting, HBD = Hydrogen bond donating.

Figure 1 shows how the molecular parameters change when water is added to the polyglycols. When 10 mass % water is added to PEG 300, all the molecular parameters increase slightly. When water is added to PPG 425, a significant increase in polarity occurs, but only a slight increase is seen in the hydrogen bond-donating and hydrogen bond-accepting parameters.

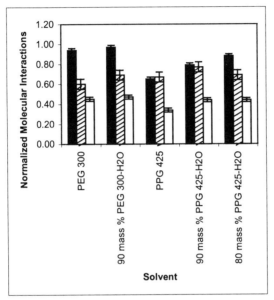

Figure 1. Effect of water on the polyglycol molecular interaction parameters. ■ *= Polarity ,* ▨ *= Hydrogen bond-accepting,* □ *= Hydrogen bond-donating.*

Viscosity

Viscosity measurements for the polyglycol-H_2O solutions were either obtained from the literature or measured. The viscosity of hexaethylene glycol (molecular weight = 282 g/mol) is 84 cP at 20 C (*13*). Similarly, the viscosity of PPG 425 is 80 cP at 25 C (*14*). However, by raising the temperature, the viscosity of the polyglycols can be reduced significantly. For example, the viscosity of hexaethylene glycol is 15 cP at 60 C and 5.5 cP at 100 C (*13*). Adding water also decreases the viscosity of the solvent. The viscosity of a 70 mass % PEG 300-H_2O solution is 28 cP at 25 C and the viscosity of a 70 mass % PPG 425-H_2O solution is 30 cP at 22 C. Therefore, the polyglycol-H_2O

solutions are viscous, but increasing the temperature or increasing the water content can considerably lower the viscosity.

Enzymatic Reactions

To identify possible uses and limitations of the polyglycol-H_2O solvents, two enzymatic reactions, an esterification reaction and a hydrolysis reaction, were run.

For the esterification reaction, isoamyl alcohol was reacted with propionic acid to form isoamyl propionate and water. Esterification reactions are often used to produce pharmaceutical products and food additives (15). Therefore, since these products are used for human consumption, it is desirable to use a non-toxic solvent. The esterification reaction was also studied in order to understand the behavior of water in the PEG-H_2O and PPG-H_2O systems. Published literature states that all water is bound to the glycol molecules in PEG-H_2O solutions which contain greater than 60 mass % PEG (16). As can be seen, water is a product in an esterification reaction. Therefore, the esterification reaction was run in the polyglycol-H_2O systems to see if equilibrium conversions could be increased since the glycol molecules are reported to bind water.

For the enzymatic esterification reaction, an immobilized lipase enzyme (Lipozyme IM, manufactured by Novo Nordisk Bioindustrials Inc.) was used. The reaction was run in 8 mL, glass vials, and monitored using a Varian gas chromatograph having a FID detector. A capillary column was used to separate the reaction components.

When this reaction was conducted using a 70 mass % PEG 300-H_2O solution and a 70 mass % PPG 425-H_2O solution, equilibrium conversions were only 8 % and 10 % respectively. However, when the esterification reaction was run under similar conditions with hexane as the solvent, equilibrium conversions were about 90 % (17). Further experiments showed that the enzyme wasn't deactivated, but that this reaction was equilibrium limited. Therefore, water in the polyglycol-water solutions is not bound strongly to the glycol and is free to participate in the reaction. Using pure PEG 300 and PPG 425 would most likely increase equilibrium conversions, but solution viscosity would also increase.

For the enzymatic hydrolysis reaction, benzyl propionate was reacted with water to form benzyl alcohol and propionic acid. Experimental conditions for this reaction were similar to those for the esterification reaction. When the reaction was run in 90 mass % PEG 300-H_2O, about 85 % of the benzyl propionate was converted to benzyl alcohol. However, the amount of propionic acid produced was less than half the amount of benzyl alcohol produced. It was found that this was due to the alcohol groups from the polyethylene glycol molecules reacting with the benzyl propionate to form benzyl alcohol and other side products.

These results from the enzymatic reactions illustrate two points. First, water apparently is not strongly bound in the 70, 80, and 90 mass % polyglycol-H_2O

solutions. Because of this, hydrolysis reactions can be performed in the polyglycol-H_2O solutions. Secondly, the alcohol groups on the glycols can react to form side products which illustrates that these solvents are not inert.

Homogeneous Reactions

To better understand the advantages and limitations of the polyglycol-H_2O solvents, several homogeneous, non-catalytic reactions were performed. Specifically, pharmaceutical type reactions (a S_N1 reaction, two Diels-Alder reactions, and a S_N2 reaction) were investigated.

S_N1 Reaction

Figure 2 shows the model S_N1 reaction that was studied.

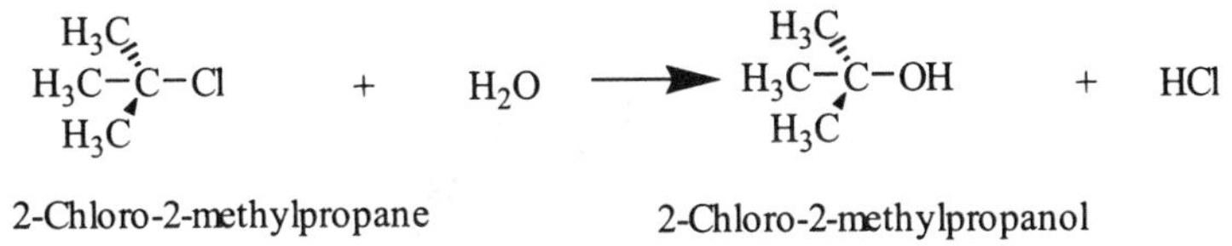

2-Chloro-2-methylpropane   2-Chloro-2-methylpropanol

Figure 2. S_N1 reaction between a tertiary halide and water to form a tertiary alcohol.

The progress of the reaction was monitored using a YSI Scientific conductivity meter. Calibration curves were made by adding known amounts of HCl to the solvents and measuring the conductance. The S_N1 reactions were run at room temperature (ca. 22 C). The reaction was run in an open vessel that was rapidly stirred. Conductivity readings were recorded until they remained constant. A gas chromatograph was used to analyze the reaction products; side reactions between the alcohol groups on polyglycol molecules and 2-chloro-2-methyl propane appeared to be negligible.

In Figure 3, the rate constants obtained using the PEG 300-H_2O mixtures are compared with the rate constants obtained from traditional solvents (*18*). (Note that the rate constants for the traditional solvents were obtained at 25 C.) It can be seen that, at low mass % of co-solvent in H_2O, the rate constants are comparable to those for the traditional organic solvents at the same mass % water. However, at high mass % co-solvent, the PEG 300-H_2O solvents result in rate constants 1-3 orders of magnitude greater. Since it may be necessary to run many S_N1 reactions at low water concentrations in order to obtain good reactant solubility, the PEG 300-H_2O system may be a good alternative solvent for S_N1 reactions.

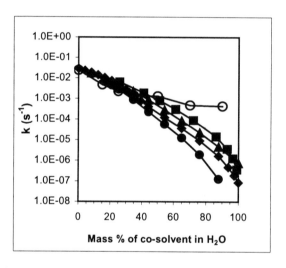

Figure 3. Rate constants for the hydrolysis of 2-chloro-2-methyl propane (T=22-25 C). ○ = *PEG 300,* ■ = *acetic acid (18)* , ▲ = *methanol (18)* , ♦ = *ethanol (18),* ● = *acetone (18).*

Diels-Alder Reactions

Figure 4 shows the two Diels-Alder reactions that were studied as model reactions.

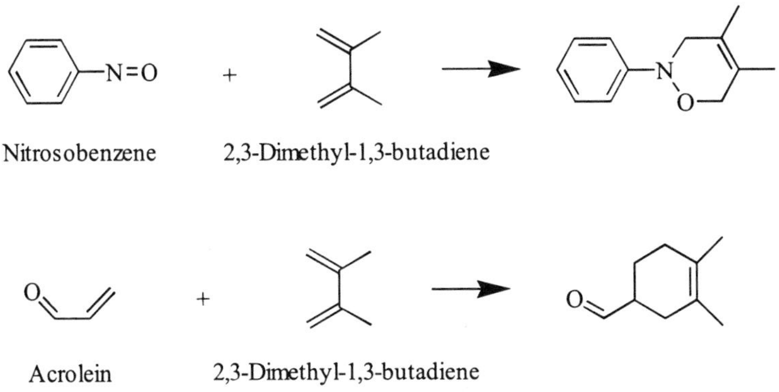

Nitrosobenzene   2,3-Dimethyl-1,3-butadiene

Acrolein   2,3-Dimethyl-1,3-butadiene

Figure 4. Diels-Alder reactions between 2,3-dimethyl-1,3-butadiene (diene) and two different dienophiles (acrolein and nitrosobenzene).

The nitrosobenzene/2-3-dimethyl-1,3-butadiene reaction was monitored by making in-situ measurements using an UV-Vis spectrophotometer (HP model 8452A). These in-situ measurements were carried out in a 3.0 mL, 45 mm, standard silica cell with a PTFE cover to hinder evaporation of the solvent or reactants.

The rate constants for the Diels-Alder reaction of nitrosobenzene and 2,3-diemethylbutadiene in PEG 300, PPG 425, and two traditional organic solvents, methanol and dichloromethane, are shown in Figure 5. It is seen that rate constants obtained in PEG 300 and PPG 425 are higher than those obtained in the traditional solvents; the rate constant obtained in PEG 300 is 2.5 times larger than the rate constant in methanol and 3.3 times larger than the rate constant in dichloromethane.

It has been shown that the addition of water to a Diels-Alder reaction can lead to rate constant increases. These increases have been attributed to both a hydrophobic effect and hydrogen bonding (*19, 20, 21, 22*). When 10 mass % water was added to PPG 425, PEG 300, and methanol, the rate constants increased by 40%, 80%, and 100% respectively.

The acrolein and 2,3-dimethyl-1,3-butadiene reaction, studied by Clontz (*23*), was monitored using a Hewlett Packard model 5890A gas chromatograph with a thermal conductivity detector. All reactions between acrolein and 2,3-dimethyl-1,3-butadiene were carried out in capped vials, using magnetic stir bars for agitation. Rate data were obtained by monitoring the disappearance of 2,3-dimethyl-1,3-butadiene over time.

The rate constants for the Diels-Alder reaction between acrolein and 2,3-dimethyl-1,3-butadiene are shown in Figure 6. The rate constant obtained in 70

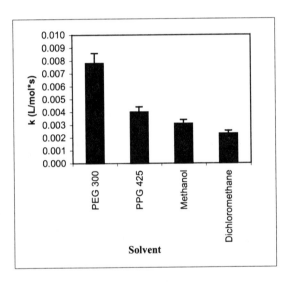

Figure 5. Diels-Alder rate constants for the reaction between 2,3-dimethyl-1,3-butadiene and nitrosobenzene (T=22 C).

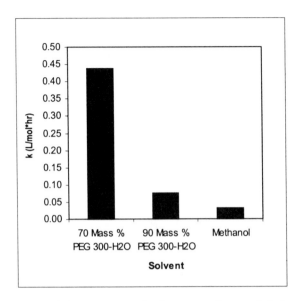

Figure 6. Diels-Alder rate constants for the reaction between 2,3-dimethyl-1,3-butadiene and acrolein @ T=22 C (23).

mass % PEG 300-H$_2$O solvents is 14 times greater than that obtained in methanol. Thus, both Diels-Alder reactions have larger rate constants in the PEG 300-H$_2$O solvents than in the traditional organic solvents. It can also again be seen that increasing the water content in the solvent increases the rate constant considerably. Higher water concentrations weren't tested because of low reactant solubility.

S$_N$2 Reaction

For a model S$_N$2 reaction, diethylamine was reacted with 1-bromobutane to form diethylbutylamine and HBr. A Varian Model 3350 gas chromatograph was used to monitor the progress of the S$_N$2 reaction, using a Varian 1075 Split/Splitless capillary injector (run using the split mode) and a FID detector. All reactions were run at room temperature in a Kontes, glass, 50 mL, jacketed reaction beaker. When obtaining rate constants for the diethylamine/bromobutane reaction, only initial rate data were used.

Rate constants for the S$_N$2 reaction in pure PEG 300, PPG 425, and typical S$_N$2 organic solvents, DMSO, DMF, acetonitrile, and ethyl acetate, are shown in Figure 7. It can be seen that except for ethyl acetate, the typical organic solvents have higher rate constants than the polyglycols tested. Modeling of a similar S$_N$2 reaction by use of the solvatochromic parameters (9,10,11,12) has shown that as the polarity of the solvent increases, the rate constant increases (*24*). From their study, it was found that hydrogen bonding doesn't seem to have an effect on the rate constants. These predictions seem to apply well to the traditional organic solvents, but fail to explain the results found for the polyglycol-H_2O solvents. Attempts are now being made to correlate the polyglygol-H_2O solvent effects on such reaction rates.

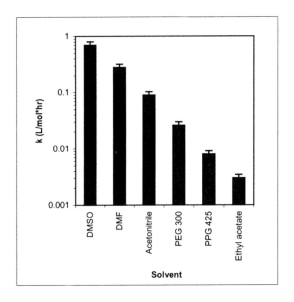

Figure 7. S_N2 rate constant comparison in six different solvents (T = 22 C).

When 10-15 mass % water is added to PEG 300 and PPG 425, the rate constants nearly double. However, when a similar mass % water is added to DMF and acetonitrile, the rate constants decrease by 10-20 %, making the alternative solvent rate constants more comparable with the traditional organic solvents.

Even though the traditional S_N2 solvents result in higher rate constants for the reaction, the polyglycol-H_2O solvents may be viable alternative solvents because of environmental and toxicity concerns. Both acetonitrile and DMF are listed as Hazardous Air Pollutants and are regulated under the Clean Air Act (*25*). DMF is a suspected carcinogen, acetonitrile is quite toxic, and both acetonitrile and DMF have shown reproductive effects (*8*).

Solubility Results

A favorable aspect of polyglycol-H_2O solvents is that organic compounds appear to be quite soluble in these alternative solvents. Table 2 shows the solubility of hexane (a non-polar, hydrophobic compound), 2-octanone (a slightly polar compound), and heptanol (a moderately polar compound) in 70 mass % PEG 300-H_2O and 70 mass % PPG 425-H_2O (23). It can be seen that hexane has a low solubility in the 70 mass % PEG-H_2O solution, but the solubility is higher in the 70 mass % PPG 425-H_2O solution. 2-octanone is slightly soluble in both the 70 mass % PEG 300-H_2O and 70 mass % PPG 425-H_2O solutions, and heptanol is quite soluble in both polyglycol-H_2O solutions.

Table II. Organic compound solubilities @ T=22 C (23) .

Solvent	Hexane, mol/L	2-Octanone, mol/L	Heptanol, mol/L
70 % PEG 300-H_2O	0.01	0.1	0.8
70 % PPG 425-H_2O	0.2	0.4	0.8

By decreasing the water content in the solvent, the solubility of non-polar, hydrophobic compounds can be increased in the polyglycol-H_2O solvents. Table 3 shows that two fairly hydrophobic compounds, 2,3-dimethyl-1,3-butadiene and 1-bromobutane, are quite soluble in pure PEG 300 and PPG 425 solutions.

Table III. Organic compound solubilities @ T=22 C.

Solvent	DMB, mol/L	1-Bromobutane, mol/L
PEG 300	1.7	4.5
90 % PEG 300-H_2O	0.5	1.2
PPG 425	completely miscible	completely miscible
90 % PPG 425-H_2O	2.6	2.8

NOTE: DMB = 2,3-Dimethyl-1,3-butadiene

When 10 mass % water is added to the solvents, the solubility decreases, but still remains significant. These results also again show that hydrophobic compounds have higher solubilities in PPG 425 solutions than in PEG 300 solutions. It is also important to note that 2,3-dimetheyl-1,3-butadiene is a common diene used for Diels-Alder reactions. Also, haloalkanes, like 1-bromobutane, are used in many S_N2 reactions. Therefore, these solubility results combined with the rate constant results, show that significant overall reaction rates can be achieved using these alternative solvents.

Conclusions

PEG 300-H_2O, PPG 425-H_2O, as well as pure PEG 300 and PPG 425 are non-volatile, environmentally benign solvents. In this work, these solvents have been investigated as possible alternative solvents for fine chemical synthesis. Possible uses and limitations have been identified.

It was found that these solvents are reasonably polar solvents and have hydrogen bond-donating and hydrogen bond-accepting capabilities. Although the polyglycol-H_2O solutions are also viscous, their viscosity can be greatly reduced by increasing the temperature or by increasing their water content.

Two enzymatic reactions using the polyglycol-H_2O solvents showed that water in the polyglycol-H_2O solutions is not tightly bound and is free to react. These results showed that the alcohol groups on the glycols can also react.

Four, non-catalytic, homogeneous reactions were run using the polyglycol-H_2O solvents. At high mass % co-solvent, the rate constants for the S_N1 reaction obtained in PEG 300-H_2O were 1-3 orders of magnitude higher than those obtained in the traditional organic-H_2O solvents (at the same mass % co-solvent). The rate constants of two Diels-Alder reactions in the polyglycol-H_2O solvents were larger than those obtained in two common organic solvents, methanol and dichloromethane. A S_N2 reaction was also run these solvents. Even though the S_N2 rate constants obtained in the polyglycol solutions were lower than the rate constants in the traditional organic solvents, the alternative solvents still may be good replacement solvents because of environmental and toxicity issues.

Organic compound solubilities were obtained in the polyglycol-H_2O solutions. In the 70 mass % PEG 300-H_2O and 70 mass % PPG 425-H_2O solutions, heptanol was quite soluble while hexane was relatively insoluble. However, in pure PEG 300, pure PPG 425, 90 mass % PEG-H_2O, and 90 mass % PPG 425-H_2O, even non-polar, hydrophobic compounds had significant solubilities.

From the rate constant results and the solubility results, it can be seen that significant overall reaction rates can be achieved using these solvents depending on the specific reaction. Also, it appears that the amount of water affects both the reaction rate constants and reactant solubilities.

The feasibility of using these solvents also depends on the ability to recover both the products and the polyglycols from the resulting solution. However, it would appear that separation of the products from the polyglycol solutions should be quite easy since organic compounds are not soluble in these solutions at high water contents. Therefore, simple addition of water to the final solution will result in the formation of two phases, one containing the product and the other containing the polyglycol and water. Separation of the polyglycols from the resulting aqueous solution could then be done using vacuum distillation or some other method. These aspects will be considered in future work.

The use of polyglycol solutions such as PEG 300, PPG 425, and their aqueous solutions, appear to offer promise as alternative, environmentally benign solvents for selected chemical reactions. More research with other types of reactions is needed to fully assess the applicability of these systems; however, the polyglycol solutions seem to be good alternatives for several pharmaceutical-type reactions.

Acknowledgements

This work was supported by NSF Training Grant DGE9452654 and by a grant from the U.S. Environmental Protection Agency's Technology for Sustainable Environment (TSE) program. Although the research described in the article has been funded in part by the U.S. Environmental Protection Agency's TSE program through grant R82813301-0, it has not been subjected to any EPA review and therefore does not necessarily reflect the views of the Agency, and no official endorsement should be inferred.

References

1. Sherman, J.; Chin, B.; Huibers, P. D.; Garcia-Valls, R.; Hattan, T. A. Solvent Replacement for Green Processing. *Environ. Health Perspect.*, **1998**, *106*, 253-271.
2. Kirschner, E. M. Environment, Health Concerns Force Shift In Use Of Organic Solvents. *Chem. Eng. News*, **1994,** *72*, 13-20.
3. Anastas, P. T.; Benign By Design Chemistry. In *Benign by Design: Alternative Synthetic Design for Pollution Prevention*; Anastas, P. T.; Farris, C.A., Eds.; ACS Symposium Series 577; American Chemical Society: Washington, DC, 1994; Chapter 1, pp 2-23.
4. *1997 Toxics Release Inventory: Public Release Data;* EPA 745-C-98-004; U.S. Environmental Protection Agency: Washington DC, 1997.
5. *Profile of The Pharmaceutical Industry;* EPA 310-R-97-005; U.S. Environmental Protection Agency, Office of Compliance, U.S. Government Printing Office: Washington DC, 1997.

6. Zalipsky, S.; Harris, J. M.; Introduction to Chemistry and Biological Applications of Poly(ethylene glycol). In *Poly(ethylene glycol): Chemistry and Biological Applications*; Zalipsky, S., Harris, J. M., Eds.; American Chemical Society: Washington DC, 1997; Chapter 1, pp 1-13.

7. *The Merck Index*, 12th ed.; Budavari, S. Ed.; Merck and Co.: Rahway, NJ, 1996; p 7733.

8. Lewis, R. J., Sr. *SAX'S Dangerous Properties of Industrial Materials*; 10th ed.; Wiley and Sons: New York, NY, 2000.

9. Kamlet, J. M.; Abboud, J. M.; Abraham, M. H.; Taft, R. W. Linear Solvation Energy Relationships. 23. A Comprehensive Collection of the Solvatochromic Parameters, π^*, α, and β, and Some Methods for Simplifying the Generalized Solvatochromic Equation. *J. Org. Chem.*, **1983**, *48*, 2877-2887.

10. Kamlet, J. M.; Taft, R. W. The Solvatochromic Comparison Method. 1. The β-Scale of Solvent Hydrogen-Bond Acceptor (HBA) Basicities. *J. Am. Chem. Soc.*, **1976**, *98*, 377-383.

11. Kamlet, J. M.; Abboud, J. M.; Taft, R. W. The Solvatochromic Comparison Method. 6. The π^* Scale of Solvent Polarities. *J. Am. Chem. Soc.*, **1977**, *99*, 6027-6038.

12. Taft, R. W.; Kamlet, M. J. The Solvatochromic Comparison Method. 2. The α-Scale of Solvent Hydrogen-Bond Donor (HBD) Acidities. *J. Am. Chem. Soc.*, **1976**, *98*, 2886-2894.

13. Lee, R.; Teja, A. S. Viscosities of Poly(ethylene glycols). *J. Chem. Eng. Data*, **1990**, *35*, 385-387.

14. *Aldrich Handbook of Fine Chemicals and Laboratory Equipment*, Sigma-Aldrich Co.: USA, 2000; p 1387.

15. Esters, Organic. *Ulmann's Encyclopedia of Chemical Technology*, 5th ed.; VCH Publishers: Weinheim, Germany, 1987; Vol. A9, pp 565-585.

16. Tilcock, C. P. S.; Fisher, D. The Interaction of Phospholipid Membranes With Poly(Ethylene Glycol) Vesicle Aggregation And Lipid Exchange *Biochim. Biophys. Acta*, **1982**, *688*, 645-652.

17. Mensah, P.; Gainer, J. L.; Carta, G. Adsorptive Control of Water in Esterification with Immobilized Enzymes: I. Batch Reactor Behavior. *Biotechnol. Bioeng.*, **1998**, *60*, 434-444.

18. Fainberg, A. H.; Winstein, S. Correlation of Solvolysis Rates. III. t-Butyl Chloride in a Wide Range of Solvent Mixtures. *J. Am. Chem. Soc.*, **1956**, *78*, 2770 – 2777.

19. Otto, S.; Engberts, J. B. F. N. Diels –Alder Reactions in Water. *Pure Appl. Chem.*, **2000**, *72*, 1365-1372.

20. van der Wel, G. K.; Wijnen, J. W.; Engberts, J. B. F. N. Solvent Effects on a Diels-Alder Reaction Involving a Cationic Diene: Consequences of the

Absence of Hydrogen-Bond Interactions for Accelerations in Aqueous Media. *J. Org. Chem.*, **1996**, *61*, 9001-9005.

21. Cativiela, C.; Garcia, J. I.; Mayoral, J. A.; Salvatella, L. Modeling of Solvent Effects on the Diels-Alder Reaction. *Chem. Soc. Rev.*, **1996**, *25*, 209.

22. Otto, S.; Blokzijl, W.; Engberts, J. B. F. N. Diels-Alder Reactions in Water. Effects of Hydrophobicity and Hydrogen Bonding. *J. Org. Chem.*, **1994,** *59*, 5372-5376.

23. Clontz, R. Polyglycol Solutions as Alternatives to Organic Solvents. M.S. Thesis, University of Virginia, Charlottesville, VA, 1997.

24. Abraham, M. H. Solvent Effects on Reaction Rates. *Pure Appl. Chem.,* **1985,** *57*, 1055-1064.

25. Clean Air Act Amendments of 1990, Section 112, 1990.

Chapter 16

Greener Solvent Selection under Uncertainty

Ki-Joo Kim[1], Urmila M. Diwekar[1,*], and Kevin G. Joback[2]

[1]Department of Civil and Environmental Engineering, Carnegie Mellon University,
5000 Forbes Avenue, Pittsburgh, PA 15213
[2]Molecular Knowledge Systems, P.O. Box 10755, Bedford, NH 03110–0755

Solvent selection is an important step in process synthesis, design, or process modification. Computer-aided molecular design (CAMD) approach based on the reverse use of group contribution method provides a promising tool for solvent selection. However, uncertainties inherent in these techniques and associated models are often neglected. This paper presents a new approach to solvent selection under uncertainty. A case study of acetic acid extraction demonstrates the usefulness of this approach to obtain robust decisions.

Introduction

Solvents are extensively used as process materials (e.g., extracting agent) or process fluids (e.g., CFC) in chemical process industries, pharmaceutical industries, and solvent-based industries such as coating and painting. Since waste solvents are a main source of pollution to air, water, and soil, it is desirable to use reduced amount of solvents and/or environmentally friendly solvents without sacrificing process performance. There are some solvents which must be eliminated due to environmental and health effects and regulatory requirements. For example, the Montreal protocol bans many chlorinated solvents (*1*).

Solvent selection, an approach to generate candidate solvents having desirable properties, can help to handle these problems. Several methodologies have been developed for solvent selection over the years (*2*). First approach uses traditional laboratory synthesis and test methodology to find promising solvents. This method can provide reliable and accurate result, but in many cases this method can not be applied due to cost, safety, and time constraints. Second approach is to search the property database. Although it is the most common and simple method, it is limited by the size and accuracy of the database. Furthermore, these two methods may not provide the best solvent because there are huge number of solvent molecules to be tested or searched. Finally, computer-aided molecular design (CAMD) can automatically generate promising solvents from their building blocks or groups (*3,4*). This method can generate lists of candidate solvents with reasonable accuracy within moderate time scale. CAMD can also be applied to CFC substituents (*4*), solvent blend design (*5*), polymer and drug design (*6*), and alternative process fluid design (*7*). However, CAMD is limited by the availability and reliability of the property estimation methods.

All methodologies for solvent selection are exposed to uncertainties that arise from experimental errors, imperfect theories or models and their parameters, improper knowledge or ignorance of systems, and inadequate controls. Although uncertainties can affect the real implementation of selected solvents, few papers in literature have focused on uncertainties. In this study a new CAMD method for solvent selection under uncertainty is presented and is applied to generate greener solvents for acetic acid extraction from water as a case study.

As the ranking or priority for solvent selection in this study is based on the Hansen's solubility parameters, we describe solubility parameters and solvent selection model first. Then several sections are devoted to explain and discuss this new CAMD method, uncertainty quantification, and case study. Finally summary is followed.

Hansen's Solubility Parameter and Solvent Selection Model

The solubility parameter, δ, is one of the most important parameters in physical chemistry and thermodynamics of solutions. It can serve as a key parameter for solvent selection, solubility estimation, and the estimation of polymer swelling (8). Though it was originally introduced by Hildebrand and Scott (9), the most common one is Hansen's three-dimensional solubility parameter (10) which is given by:

$$\delta^2 = \delta_d^2 + \delta_p^2 + \delta_h^2 \quad (\text{unit}:\text{MPa}) \tag{1}$$

where δ_d is the dispersive term, δ_p is the polar term, and δ_h is the hydrogen bonding term. The solubility parameter (δ) and its three terms (δ_d, δ_p, and δ_h) can be determined by semi-empirical methods and are tabulated by Barton (11) for most common liquids.

Miscibility of two liquids A and B depend on the heat of mixing ΔH_{mix}, and ΔH_{mix} in the Hansen theory is given by the following equation.

$$\Delta H_{mix} = (n_A V_A + n_B V_B)\left\{(\delta_d^A - \delta_d^B)^2 + (\delta_p^A - \delta_p^B)^2 + (\delta_h^A - \delta_h^B)^2\right\}\Phi_1\Phi_2 \tag{2}$$

where n is the number of moles, V is the molar volume, and Φ is the volume fraction. When the heat of mixing approaches to zero, two liquids A and B are soluble or miscible with each other. Hence the three solubility parameter terms should be close to minimize the heat of mixing.

Generally speaking, solubility parameters for solvent selection are not as accurate as other property estimation methods such as infinite dilution activity coefficients (γ^∞). But solubility parameter method is universal and simple to apply, and thus can be used for guiding and screening candidate solvents with relatively acceptable accuracy.

To replace the current solvent or design new solvents, there are several criteria are used. These are (a) distribution coefficient (m), (b) solvent selectivity (β), (c) solvent loss (S_L), (d) physical properties such as boiling point, flash point, density, and viscosity, (e) chemical stability, (f) toxicology, and (g) cost. For extraction process, the final selection of solvents will generally be dominated between m and β. Distribution coefficient (m), a measure of solvent capacity, is defined as:

$$m \propto \left(\frac{r_{UR}}{r_{US}}\right)^2 \frac{MW_R}{MW_S} \tag{3}$$

where R, U, and S are raffinate, solute, and solvent, respectively. MW is molecular weight. r_{UR} and r_{US} are Euclidean distance metric between two molecules and are defined as follows:

$$r_{\text{UR}} = \left[\left(\delta_d^U - \delta_d^R \right)^2 + \left(\delta_p^U - \delta_p^R \right)^2 + \left(\delta_h^U - \delta_h^R \right)^2 \right]^{1/2}$$

$$r_{\text{US}} = \left[\left(\delta_d^U - \delta_d^S \right)^2 + \left(\delta_p^U - \delta_p^S \right)^2 + \left(\delta_h^U - \delta_h^S \right)^2 \right]^{1/2} \tag{4}$$

Liquids having similar solubility parameters are soluble or miscible with each other. The distance between solute and solvent (r_{US}) should be small while the distance between solute and raffinate (r_{UR}) is fixed for a given solute-raffinate system. Thus solvents having smaller r_{US} can increase m, and high m reduces the size of an extracting equipment and the amount of recycling solvent.

Solvent selectivity (β), the ability of the solvent to selectively dissolve solute, is the ratio between distribution coefficients of solute and raffinate, and defined by:

$$\beta = \frac{m_U}{m_R} \propto \left(\frac{r_{\text{RS}}}{r_{\text{US}}} \right)^2 \frac{\text{MW}_U}{\text{MW}_R} \tag{5}$$

where r_{RS} is defined in a similar way. High β reduces the cost of solute recovery as solvent is highly selective to solute. Solvent loss (S_L) can be expressed by the following equation:

$$S_L \propto \left(\frac{1}{r_{\text{SR}}} \right)^2 \frac{\text{MW}_S}{\text{MW}_R} \tag{6}$$

Low S_L means high selectivity toward solute and determines immiscibility between solvent and raffinate.

Table 1 shows the Hansen's three-dimensional solubility parameters and estimated solvent properties (m, β, and S_L). In this example, acetic acid is a solute, water is a raffinate, and ethyl acetate is a current solvent. The aim of solvent selection is to generate solvents having better solvent properties than the current solvent.

Table 1. The Hansen's solubility parameters and solvent properties.

Solvent	δ_d	δ_p	δ_h	Properties	Values
Acetic acid	13.9	12.2	18.9	m	1.043
Water	12.2	22.8	40.4	β	35.20
Ethyl acetate	13.4	8.60	8.90	S_L	0.0041

Computer Aided Molecular Design Under Uncertainty

Group contribution method is a forward problem; if we know a molecule, we can estimate its physical, chemical, biological, and health effect properties based on its groups or building blocks (Figure 1a). However, CAMD is a backward problem; if we know desirable properties or regulation limits, we can find molecules that satisfy these properties or limits by constructing groups (Figure 1b). CAMD approach, though not as accurate as experimentation, can generally provide satisfactory results from large scale combinations of groups.

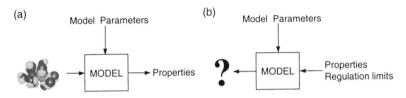

Figure 1. Forward(a) and backward(b) problems.

In group contribution methods groups or building blocks are uniquely designed to generate any possible solvent molecules, and properties of each group can be theoretically calculated, experimentally obtained, or statistically regressed. From a set of groups, all possible combinations of groups can be made to generate molecules. Once molecules are generated, desired properties of the molecules are predicted based on the properties of their groups and tested if they satisfy the pre-specified criteria. If the generated-and-tested molecules have desired properties, they are stored and sorted according to predetermined priorities. This common CAMD approach is called a generation-and-test approach (*3,4*).

Besides the generation-and-test method, mathematical optimization methods (*6,12-14*) are usually applied to solve this reverse problem. However, the optimization methods used in these earlier studies are not designed to include uncertainties which can be a major computational bottleneck for the large-scale stochastic combinatorial optimization problems. Here in this paper we are using a combinatorial optimization algorithm specially designed to efficiently handle combinatorial optimization problems under uncertainty (*15,16*). Figure 2 shows a simple representation of this approach. This approach involves two recursive loops; the outer optimization loop and the inner sampling loop. In this approach the optimizer not only determines the real decision variables like the number of groups in a solvent molecule, group indexes, but also the number of uncertain

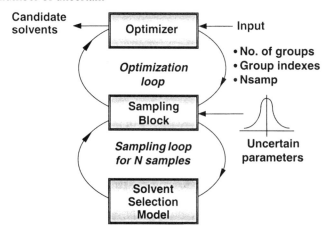

Figure 2. An optimization framework for solvent selection under uncertainty.

samples needed for the inner sampling loop. The inner sampling loop essentially converts the deterministic group contribution model into a stochastic model. Stochastic modeling involves four steps: (a) identifying and specifying key input parameter uncertainties in terms of probabilistic distribution functions, (b) sampling these distributions in an iterative fashion, (c) propagating the effects of these uncertainties through the model, and (d) analyzing the output using statistical techniques.

Total 17 groups and their properties used in this study are shown in Table 2 for estimating solubility parameters. One of the authors (Joback) determined the three solubility parameter terms using the least square method from the literature data (*11*). In this table, each column consists of group specific solubility parameters and intercept values. The solubility parameter can be estimated by linearly adding group properties. For example, ethyl acetate (CH_3-COO-CH_2-CH_3) has three distinctive groups, and its dispersive solubility parameter term is estimated to 13.38 $MPa^{1/2}$ (2x0.344+0.268-0.862+13.290). Its polar and hydrogen bonding terms are similarly 8.24 and 8.95 $MPa^{1/2}$, respectively. The literature values for δ_d, δ_p, and δ_h are 13.40, 8.60, and 8.90, respectively, and we can see that estimated and literature data are very close.

The set of groups in Table 2 is specially designed for linear or branched hydrocarbons while aromatic, cyclic, and/or halogenated compounds are eliminated due to environmental concerns. As described earlier, one of the features of solubility parameter method is universality: we can use the same

Table 2. Solubility parameters of groups.

Groups	Dispersive	Polar	Hydrogen-bonding
-CH$_3$	0.344	-0.591	-0.848
-CH$_2$-	0.268	-0.377	-0.595
>CH-	-0.142	-0.801	-1.172
>C<	-1.163	-1.039	-2.496
CH$_2$=CH-	-1.163	-1.039	-2.496
CH$_2$=C<	-0.243	0.275	-3.542
-CH=CH-	-0.566	-0.034	-0.0776
-CH=C<	-0.695	-0.529	-3.175
>C=C<	-0.823	-1.025	-5.574
-OH	-0.648	5.548	10.630
-O-	-0.638	2.315	1.804
>C=O	-1.145	4.670	4.486
O=CH-	-1.114	5.922	5.256
-COOH	1.068	6.942	11.120
-COO-	-0.862	4.729	4.012
>NH	-1.074	3.875	2.772
-CN	-1.628	6.904	8.317
Intercept	13.290	5.067	7.229

groups for estimating other properties such as boiling point (*17*), and thus there is no need to have another group set.

4. Uncertainty Identification and Quantification

The Hansen solubility parameters of liquid molecules are estimated by semi-empirical methods, and the three solubility parameter terms for *each group* are regressed using the least square method. Table 3 shows some estimation errors of the Hansen's three-dimensional solubility parameters. For example, ethanol is showing 15% relative error in the solubility parameter mainly due to the error in the δ_h term. Estimated solubility parameters of ethyl acetate are quite close to the literature values, which is also shown in Table 1. For diisobutyl ketone, we can find an interesting result. Though each solubility parameter term has large discrepancy, the resulting total solubility parameter is closer to the value reported in the literature. What is important in this study is to quantify uncertainty in each solubility parameter term, not the uncertainty in total solubility parameter, that governs miscibility of two liquids as shown in eq 2.

Table 3. Example calculations of Hansen's solubility parameters.

	Ethanol			Ethyl acetate			Diisobutyl ketone		
	Lit.	Est.	Δ(%)	Lit.	Est.	Δ(%)	Lit.	Est.	Δ(%)
δ_d	12.6	11.7	-7.4	13.4	13.4	-0.1	14.5	13.8	-5.0
δ_p	11.2	9.7	-13.9	8.6	8.2	-4.2	6.8	5.0	-26.2
δ_h	20.0	16.4	-17.9	8.9	9.0	0.6	3.9	5.2	32.0
δ	26.2	22.3	-14.5	18.2	18.1	-0.9	15.5	15.5	-5.8

The group contribution method in this study has 17 groups as shown in Table 2 and each group has three solubility parameter terms. Since the total uncertainty distributions are 51, it is impractical and statistically insignificant to figure out each uncertainty distribution. Instead, the uncertainties of the Hansen solubility parameters (δ_d, δ_p, and δ_h) of a liquid molecule are analyzed and quantified in terms of a new parameter called Uncertainty Factor (UF). We define Uncertainty Factor (UF) as the ratio of the literature solubility parameter (11) to the estimated solubility parameter using the group contribution method.

$$UF = \frac{\delta_{lit}}{\delta_{est}} \times 100 \quad (\%) \qquad (7)$$

where the UF can be applied to dispersive, polar, or hydrogen bonding terms. Note that the UF of 100 % means that the estimated value is exactly same to the literature value.

In order to elicit the UFs, the estimated solubility parameters of 66 non-cyclic and non-aromatic compounds are compared with the literature values. The probabilistic distributions of the three UFs associated with the three solubility parameters are shown in Figure 3. The UF of the dispersive term (δ_d) is normally distributed with 105.4 % mean and 8.3 % standard deviation. The UFs of δ_p and δ_h are normally distributed with 121.4 % mean and 128.1 % standard deviation

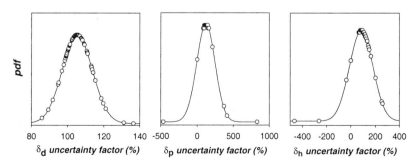

Figure 3. The UFs for dispersive, polar, and hydrogen bonding terms.

and 84.5 % mean and 95.5 % standard deviation, respectively. From this figure, we can see that the effect of uncertainty on δ_d is not significant as compared to the effects on the polar term (δ_p) and the hydrogen bonding term (δ_h). Therefore, we can say that UF increases estimated δ_p and decreases estimated δ_h, and the resulting δ, m, β, and S_L are subject to change.

Case Study: Solvent Selection for HOAc Extraction

Acetic acid is commonly used as a process solvent or is produced as a byproduct. Therefore, acetic acid is a pollutant as well as a valuable solvent, and it is desirable to minimize the discharge of acetic acid to the environment. To recycle or remove acetic acid from process streams or process units, extraction process is generally applied. For extraction, we can either use high boiling solvents (18,19) or low boiling solvents (4). Ethyl acetate, isoamyl acetate, and isopropyl acetate are widely used in industries to extract acetic acid. This study is mainly focused on finding candidate solvents having low boiling points.

The generation-and-test CAMD approach is usually computer-intensive as this method tries to generate all possible molecular combinations. If the maximum number of groups in a molecule, for example, is 12, then the total number of possible molecular combinations are 17^{12} (5.8×10^{14}). In addition, if we consider uncertainties on the solubility parameter terms of groups, this problem becomes computationally expensive.

In order to reduce computational burden and guarantee best candidate solvents, a discrete stochastic optimization method is implemented (15,16) as shown in Figure 2. The optimizer in the upper loop determines the number of groups (NG) in a solvent molecule, group indexes ($N(i)$, $i=1, \ldots, NG$) that tells which groups are present in a proposed solvent, and the number of uncertain samples (N_{samp}). The information about the distribution functions in terms of 0.1% and 99.9% quantiles is supplied to the inner sampling loop. This sampling loop then uses an efficient sampling method (20) to generate the uncertain samples. Each sample is propagated through the model based on the group contribution method to evaluate the expected values of distribution coefficient, solvent selectivity, and boiling points of each component. This information is transferred to the optimization loop as the objective function and constraints. The optimizer determines whether or not the probabilistic results of the given molecule are optimal.

The main priority (objective function) of solvent selection is distribution coefficient, while the other properties are used as constraints that are summarized in Table 4. The number of groups (NG) in a solvent molecule spans

Table 4. Experimental conditions for solvent selection under uncertainty.

Parameter	Bounds
No. of groups in a solvent (NG)	2 ~ 12
β (solvent selectivity)	$\beta \geq 17.43$
S_L (solvent loss)	$S_L \leq 0.0045$
Boiling point ($^{\circ}$C)	47 ~ 108

from 2 to 12. The bounds of β, S_L, and boiling points are based on the values of ethyl acetate that is one of common solvents for acetic acid extraction.

Table 5 shows the top 20 candidate solvents of the deterministic and stochastic cases. For the deterministic case, the first 10 solvents are alcohols and the remaining solvents are mostly aldehyde. Some of the promising solvents in this case are ethyl alcohol (No. 1), propyl alcohol (No. 3), isopropyl alcohol (No. 5), acetone (No. 12), and methyl ethyl ketone (No. 16). The alcohol function group has the largest hydrogen-bonding term and the second largest

Table 5. Top 20 candidate solvents.

	Deterministic case			Stochastic case	
No	Solvent	m	No	Solvent	m
1	CH_3,CH_2,OH	17.2	1	$3CH_3,CH,COO$	161
2	$CH_3,CH=CH,OH$	12.6	2	$2CH_3,CH_2=C,CO$	20.1
3	$CH_3,2CH_2,OH$	9.50	3	$2CH_3,CH,OH$	5.95
4	$CH_2=CH,OH$	9.34	4	$CH_3,CH_2,CH_2=CH,COO$	4.91
5	$2CH_3,CH,OH$	6.06	5	$2CH_3,COO$	4.26
6	$CH_2=CH,2CH_2,OH$	5.83	6	$CH_3,2CH_2,OH$	3.90
7	$CH_3,CH_2=C,OH$	5.20	7	CH_3,CH_2,OH	3.53
8	$CH_2=CH,2CH_2,OH$	3.86	8	$2CH_3,CO$	3.51
9	$CH_3,CH_2=C,CH_2,OH$	3.41	9	$2CH_2,CH_2=CH,OH$	3.38
10	$CH_2,CH_2=CH,CH,OH$	2.78	10	$CH_2,CH2=CH,OH$	3.34
11	$CH_2,CH=CH,CHO$	2.07	11	$2CH_3,CH=C,CHO$	2.85
12	$2CH_3,CO$	2.06	12	$CH_2=CH,OH$	2.77
13	$CH_3,2CH_2,CHO$	1.83	13	$CH_2,CH=CH,OH$	2.58
14	$CH_3,CH_2,CH=CH,CHO$	1.50	14	$CH_2,2CH_2,CHO$	2.40
15	$2CH_3,CH,CHO$	1.47	15	$2CH_3,CH_2,CO$	2.16
16	$2CH_3,CH_2,CO$	1.44	16	$CH_3,CH_2=C,OH$	2.10
17	$CH_2=CH,CH2,CHO$	1.42	17	$2CH_3,CH_2,CH=CH,O$	2.09
18	$2CH_3,COO$	1.39	18	$2CH_3,CH=CH,CO$	2.01
19	$CH_3,3CH_2,CHO$	1.33	19	$2CH_3,CH_2=C,COO$	1.86
20	$CH_3,CH_2=C,CHO$	1.22	20	$2CH_3,CH_2,CH,CHO$	1.85

polar term. This feature of the OH group decreases r_{US}, resulting in an increase of m. Similarly the CHO group has the second largest hydrogen-bonding term and the largest polar term which also decrease r_{US}. Large difference in hydrogen-bonding terms between OH and CHO groups (10.63 vs. 5.26) and small difference in the polar terms (5.55 vs. 5.92) make alcohols preferred solvents for acetic acid extraction.

However, the stochastic case provides a different set of candidate solvents. Only 13 of the solvents generated at the deterministic case are appeared in the list of the stochastic case (See bold numbers in both columns). Some of promising solvents are isopropyl acetate (No. 1), isopropyl alcohol (No. 3), acetone (No. 5), and propyl alcohol (No. 6). Isopropyl acetate that is not listed in the top 20 solvents for the deterministic case, is one of the common industrial solvents for acetic acid extraction and is proved to be highly selective for this extraction purpose. Acetone that is appeared in both cases is reported as the best solvent for this purpose by Joback and Stephanopoulos (4). They also used the solubility parameter method and similar constraints even though their CAMD approach was the generation-and-test method. The combinatorial optimization method used in this study provides more promising solvents than acetone. Ethyl acetate, one of the common industrial solvents, is generated outside the top 20 candidate solvents at both cases since the m for ethyl acetate (1.04) is relatively low.

Because the mean value of the uncertainty factor of δ_h is 84.9%, the contribution of the hydrogen-bonding term decreases in the stochastic case and results in solvents having various functional groups. In addition, the reduced δ_h term also decreases the resulting distribution coefficients. If we look at the distribution coefficients of solvents generated in both cases, the expected value of m under uncertainty is slightly smaller than that of the deterministic case.

Figure 4 shows frequencies of top 40 candidate solvents in both cases. For the deterministic case, as expected, OH and CHO groups are the most common types in the candidate solvents. However, for the stochastic case, other functional groups also have high frequencies, and alkanes and alkenes are in the list of optimal solvents. This means that the stochastic case provide wider range of solvents.

Probabilistic density functions (pdf) of distribution coefficients for both cases are shown in Figure 5. The pdf for the deterministic case is lognormally distributed while the one for the stochastic case is governed by the Weibull distribution with the shape parameter of 1.51 and the scale parameter of 2.58. The Weibull distribution is one of the asymptotic distributions of general extreme value theory; hence, this distribution can approximate the extremely high value of m of isopropyl acetate under uncertainty. We can also conclude from this figure that the pdf of the stochastic case is narrower and showing

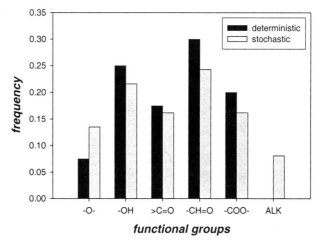

Figure 4. Frequencies of functional groups (ALK means alkanes and alkenes).

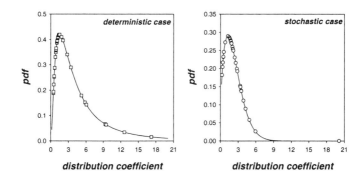

Figure 5. Probabilistic density functions of the distribution coefficient (m).

robustness in the solvent selection process. The reduced δ_p and the resulting smaller m can be attributed to this narrow peak of candidate solvents.

For real implementations of CAMD results, solvent selection and its process application (e.g., extraction process) should be simultaneously integrated. In addition, objectives such as process economics, performance, and environmental impacts should be considered. This results in a complex multiobjective optimization programming and represents the current focus for the authors.

Summary

This paper presents a new combinatorial optimization approach for CAMD under uncertainty. CAMD in this study is based on the group contribution method of the Hansen's three-dimensional solubility parameters. We have used experimental results to quantify uncertainties. A new parameter called uncertainty factor, UF is introduced to quantify uncertainties in the solubility parameter. A real world case study for solvent selection for acetic acid extraction from water is presented to illustrate the approach. For deterministic case, solvents such as ethyl alcohol, propyl alcohol, isopropyl alcohol, acetone, and methyl ethyl ketone are generated. However, under the stochastic case, a different set of solvents such as isopropyl acetate, isopropyl alcohol, acetone, and propyl alcohol are generated. The uncertainty case also identified a commonly used industrial solvent as an important solvent that was not in the top list in the deterministic studies. The analysis showed that the uncertainty results provide wider selection window and robustness in solvent selection. This combinatorial optimization method can also provide more promising solvents than the generation-and-test method.

Acknowledgement

The authors thank the National Science Foundation for their funding this research (CTS-9729074).

References

1. The Montreal Protocol on Substances that Deplete the Ozone Layer. **1987**.
2. Zhao, R.; Cabezas, H. *Ind. Eng. Chem. Res.* **1998**, 37, 3268.
3. Gani, R.; Brignole, E.A. *Fluid Phase Equilibria* **1983**, 13, 331.
4. Joback, K.G.; Stephanopoulos, G. *Proc. FOCAPD '89*, Snowmass, CO, 1989, pp 363.
5. Buxton, A.; Livingston, A.G.; Pistikopoulos, E.N. AIChE Journal, **1999**, 45, 817.
6. Vaidyanathan, R.; El-Halwagi, M. *Ind. Eng. Chem. Res.* **1996**, 35, 627.
7. Constantinou, L.; Jaksland, C.; Bagherpour, K.; Gani, R.; Bogle, I.D.L. *AIChE Symposium Series*, AIChE: New York, **1994**, Vol. 90, No. 303; pp 105.
8. Panayiotou, C. *Fluid Phase Equilibria*, **1997**, 131, 21.
9. Hildebrand, J.; Scott, R.L. *Regular Solutions*, Prentice-Hall: Englewood Cliffs, NJ, 1962.

10. Hansen, C.M. *Ind. Eng. Chem. Prod. Res. Dev.* **1969**, 8, 2.
11. Barton, A.F.M. *CRC Handbook of Solubility Parameters and Other Cohesion Parameters*, CRC Press: Boca Raton, FL, 1983, pp 94.
12. Odele, O.; Macchietto, S. *Fluid Phase Equilibria*, **1993**, 82, 47.
13. Maranas, C.D. *Ind. Eng. Chem. Res.* **1996**, 35, 3403.
14. Venkatasubramanian, V.; Chan, K.; Caruthers, J.M. *Comp. Chem. Eng.* **1994**, 18, 833.
15. Chaudhuri, P.; Diwekar, U.M. *AIChE J.* **1996**, 42, 742.
16. Chaudhuri, P.; Diwekar, U.M. *AIChE J.* **1999**, 45, 1671.
17. Joback, K.G; Reid, R.C. *Chem. Eng. Commun.* **1987**, 57, 233.
18. Pretel, E.J.; Lopez, P.A.; Bottini, S.B.; Brignole, E.A. *AIChE J.* **1994**, 40, 1349.
19. Hostrup, M.; Harper, P.M.; Gani, R. *Comp. Chem. Eng.* **1999**, 23, 1395.
20. Kalagnanam, J.R.; Diwekar, U.M. *Technometrics* **1997**, 39, 308.

Chapter 17

Dibasic Ester: A Low Risk, Green Organic Solvent Alternative

Nicholas E. Kob

DuPont Nylon Intermediates and Specialties, P.O. Box 80302, Wilmington, DE 19880
(phone: (302) 695–4774; fax: (302) 695–3817)

Over 12 billion pounds of solvents are sold globally each year and they
will continue to be valued chemicals in many market industries. However,
health and environmental regulations are increasingly driving solvent selection.
The paper presents a new tool, the vapor hazard ratio (VHR), that aids one in
selecting a safe organic solvent. Also, the paper describes Dibasic Ester (DBE)
which is a safe environmentally friendly organic solvent derived from a plant by-
product stream.

Introduction

Solvents are everywhere. There are over 60 companies engaged in
solvent manufacturing in the United States. In 1995 the global solvent market
consumed approximately 12 billion pounds of solvents valued at nearly 4 billion
dollars[1]. Demand for solvents is spread throughout the industrial, manufacturing
and consumer market segments. Solvent usage is predicted to remain at the
same levels over the next 5 years. Given these statistics, it is clear that solvents
have been and will continue to remain an essential part of the industrial,
manufacturing and consumer markets. Although overall solvent demand is flat,
there is a great deal of volume shift between classes of solvents. The primary
force driving the shift is the regulatory, environmental and health and safety
concerns surrounding the use of solvents. For example, the Montreal Protocol
resulted in the phasing out of several ozone depleting solvents. Today the Clean

Air Act has placed numerous limitations on the usage of some solvents in certain applications.

These concerns have led to a decrease in usage (-3% by volume) of certain solvent classes such as aromatics, and chlorinated solvents. In contrast, the usage of lower risk solvent classes regarded as "green" such as alcohols, and esters are on the rise (+3% by volume).

The industry today seeks solvents that offer performance while minimizing health/safety and environmental impact. Solvents meeting such criteria are labeled as "green" solvents. A better description of green solvents is low risk solvents since any chemical use has some inherent risk, and the ability to manage that risk is what makes a solvent green. The desire to work with "green" solvents is very logical and we should not detour from this path. The question then becomes how does one differentiate a solvent as "green" and another as not being "green"? A better question is how does one compare risk potentials of solvents so that the one with the lowest risk is chosen. The requirements for being "green" are not well established and that has resulted in the term being used generically to describe all kinds of solvents and technologies. Green chemistry is broadly defined as the design of chemical processes and products that reduce negative impact to human health and the environment relative to the current state of the art. This definition defines attributes, but offers no specific standards. Being considered "green" has become a market driver in the cleaning/coatings industry which has led to suppliers all trying to position their products as "green". A quick scan of any trade magazines ads reveals numerous marketing claims about how their companies solvent meets is a "green" solvent. Some of the commons marketing phrases used include:

- high flash point
- slow evaporation rate
- environmentally friendly
- worker safe
- low toxicity

The problem is that such phrases are all relative, they have no definable meaning. For example, many solvents list themselves as having a high flash point (implying they are not a fire risk), but high compared to what. Methanol (flash point 52F) has a high flash point when compared to acetone (flash point 0F), but most would still consider methanol highly flammable. The National Fire Protection Association (NFPA) has established a set of criteria for evaluating fire hazards of solvents and other materials, which is based in part on flash points. The problem is that many solvent users are not aware of such information, and assume a solvent advertised as having a high flash point is safer than one that does not advertise itself as such.

Large companies have put in place SH&E (safety, environmental and health) departments that consists of experts that are able to wade through the confusing assertions and assist in the selection of low risk solvents. Smaller to medium sized companies are many times left without guidance because they can not afford a SH&E department. Such an environment surely does not get us to the goal of having cleaning processes that utilize "green" and safe solvents. Indeed guidance is needed for many in the marketplace to fill the void between scientific fact and advertising.

Government and regulatory agencies have tried to fill that void by establishing regulations and lists to potentially guide users in solvent selection. Regulations have been developed on ozone depletion, VOC, occupational exposure limits, and flammability. All of which were derived from scientific reason with the goal environmental and worker safety, and giving the solvent user some tools to distinguish a high risk from a low risk solvent. Despite the noble intentions many in the marketplace remain unable to distinguish a high risk from a low risk solvent. One source of confusion is that the number of regulations and lists is too large and too complex for the average solvent user to comprehend. In addition, the rules and regulations have inadvertently created new hurdles for the solvents and cleaning industry that are not safety or environmentally based. Each country and/or region has independently established their own regulations and lists based on scientific data with the intentions of greater environmental and worker safety. However, in some cases the rules and regulations in one region are contradictory to what another region has established. As companies participating in the cleaning market globalize, they desire to streamline their product portfolios, specifications and technical literature. This has become increasing difficult due to regulatory contradictions. The most glaring example is the VOC regulations. The European union defines VOC as a compound with a vapor pressure greater than or equal to 10 Pascals at 293 Kelvin. In the United States, a VOC is not defined by vapor pressure rather by a reactivity method.

Green chemistry encompasses all aspects and types of chemical processes that reduce negative impact to human health and the environment relative to the current state of the art. By reducing or eliminating the use or generation of hazardous substances associated with a particular synthesis or process, chemists can greatly reduce risk to human health and the environment. To expand this definition to solvents, an ideal "green", low risk solvent would meet the following criteria:

- have low human and aquatic toxicity
- should be biodegradable, not leave an environmental footprint
- have the ability to be recycled and reused
- be a non-ozone depleting chemical
- be a non-VOC chemical
- have a high flash point (not a fire hazard)
- the production should be based on renewable raw materials and/or be derived from an already existing raw material stream

New solvent technologies have emerged as "green" alternatives, such as ionic liquids. However, these technologies will likely find limited uses in niche applications due to their physical properties and lack of significant toxicological studies. This translates into a continued need and demand for organic solvents that are truly "green". This article will focus on dibasic ester solvent, DBE, which is regarded as a "green solvent" and has found increasing application replacing high-risk solvents such as methylene chloride and aromatics. Also described is an easy to use, simple method of evaluating and comparing the potential health risk of solvents that can be used as an aid to select low risk solvents. This method is called the vapor hazard ratio or VHR and utilizes a solvents exposure limit and its physical properties to more accurately asses the risk associated with using that solvent.

Discussion

Categories of Green solvents

Solvents derived from non-petroleum based natural products represent one class of "green solvents". These solvents are not produced from predominantly petroleum based raw materials, rather their starting materials are agrochemicals. The two most common solvents made from agroproducts are soy esters and lactate esters.

Soy esters are derived from the fatty acids of soybean oil. Soybeans are the predominant oilseed crop in the world with a worldwide production of 1.4 $x10^7$ metric tons a year[2]. The oil from soybeans is rich in fatty acids such as oleic and linoleic acids. This makes soybean oil an attractive starting material for esterification to an ester solvent. Soy methyl esters have a high flash point, low toxicity, and are biodegradable.

Lactic acid is produced industrially as a fermentation by-product from the corn wet-milling industry. Esterification of lactic acid with alcohols produces lactate esters, which are used as solvents. The ethyl lactate solvent has attracted the most attention due to its excellent solvency properties. Lactate esters have low toxicity, and are biodegradable.

Solvents manufactured as a coproduct

Solvents manufactured as a coproduct of industrial chemical plants are another class of "green solvents". This category utilizes product streams other than the desired product from chemical plants to make solvents. This process results in the manufacture of solvents without having to consume new raw materials and thereby is a green process. The most recognizable product from this class of green solvents is dibasic ester or DBE solvent.

DBE is manufactured using a coproduct stream from an adipic acid manufacturing plant. 2×10^6 tons of adipic acid were manufactured worldwide in 1990. Adipic acid is reacted with hexamethylenediamine to make the polyamide nylon 6,6 that is used for apparel, carpeting, and other applications. In the process KA (cyclohexanone, cyclohexanol) is oxidized with nitric acid to produce adipic acid. In addition to nitric acid oxidation to adipic acid, oxidation of KA to glutaric and succinic acids also occurs to a much lesser extent. This results in a mixed acid coproduct stream (DBA) consisting of adipic, glutaric, and succinic acids. To utilize this stream it is reacted with methanol to make refined dimethyl esters of adipic, glutaric, and succinic acids, and the product is referred to as Dibasic Ester (DBE)[3]. The esterification process is acid catalyzed in a semi-batch operation where water is continuously removed to drive the esterification reaction. Typical conditions for the esterification are: 480 grams of DBA, 120 grams of methanol, and 5.75 grams of Witco TX catalyst are heated to 120C until the acid number drops below 10. The mixture is then cooled and distilled. The chemical process for production of DBE is shown in Figure 1. DBE is a clear, colorless liquid that has the following environmental and safety properties:

- non-HAPS (hazardous air pollutant)
- low odor
- low toxicity (LD50 oral = 8191 mg/kg) and not carcinogenic
- readily biodegradable
- high flash point (above 200F, considered not-flammable)
- recyclable through vacuum distillation
- a non-VOC in Europe
- non-ozone depleting

All of these characteristics make DBE an environmentally friendly and worker-safe alternative to more hazardous solvents. DBE can be considered a "green", low risk solvent since it meets the criteria previously described for green solvents. It is used as a solvent in a number of commercial applications such as a coating tails solvent, paint stripper, resin/polymer cleaning solvent, degreaser, and a chemical intermediate.

An example of an industry that is in need of a green, low risk solvent is the paint stripping industry. Paint stripping is a multi-million dollar a year, labor intensive industry in which there can be significant chemical exposure to workers and the environment. The traditional solvent of choice to strip paint is methylene chloride. However, it recently has come under environmental and worker safety scrutiny as a suspected carcinogen and hazardous air pollutant. This has left the paint stripping industry searching for another solvent, which is low risk, to replace methylene chloride. Dibasic ester, DBE, has solvency performance similar to that of methylene chloride for paint stripping. Hansen solubility parameters are commonly used to characterize the solvency of a solvent or solvent blend based on its polar and hydrogen bonding interactions[4]. Solubility parameters have found their use in aiding in the selection of solvents for cleaning and coatings applications. Polymers/resins/soils will dissolve in solvents with similar solubility parameters to those of the polymer/resin/soil. This is the basic principle of like dissolves like. Solubility parameters help place this principle into simple qualitative terms. Hansen parameters for a solvent are experimentally determined by testing the solvent's ability to dissolve a series of polymers whose polar and hydrogen bonding characteristics are known. The laboratory solubility testing results are entered into a computer minimization program that then determines the Hansen parameters for the solvent. The Hansen system defines three solubility parameters and relates them to Hildebrand's total or overall parameter by:

$$\delta \text{ Hildebrand} = (\delta D^2 + \delta P^2 + \delta H^2)^{1/2}$$

where δD = **Dispersive or "nonpolar" parameter**
δP = **Polar parameter**
δH = **Hydrogen bonding parameter**

The Hansen solubility parameters may then be plotted in a normal three-dimensional graph. For simplicity, the graphical representation of this

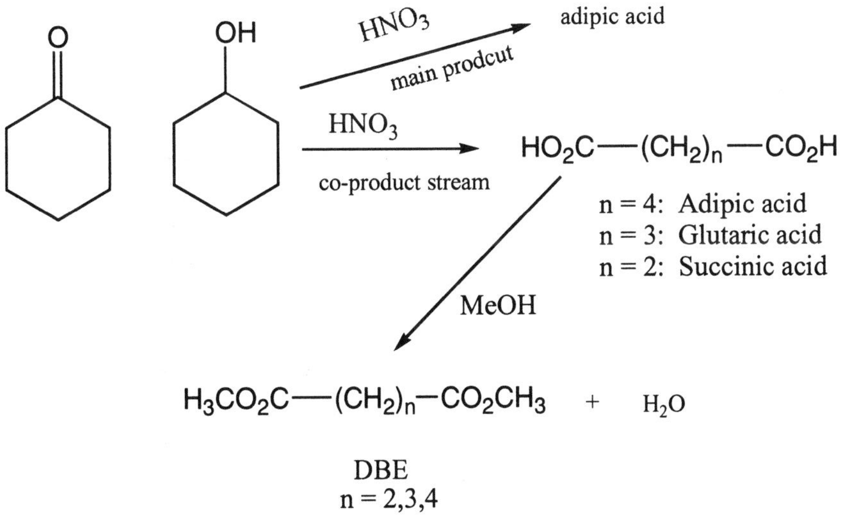

Figure 1 – DBE production

"solubility space" is often reduced to a two-dimensional plot of δP versus δH. This is generally an acceptable practice because the dispersive component parameters (δD) of many common solvents are quite similar. Using this method each solvent has its own unique place on the graph (as shown in Figure 2) which is used to describe the solvents solvency characteristics.

This Hansen theory and the computer program can also be utilized to develop an envelope of solubility for any resin/soil by placing it into a series of solvents with known Hansen parameters[4]. In practice the soil/resin is placed into 34 solvents and allowed to gently stir for 2 days. After which time the vials are checked to ascertain which solvents dissolved the soil/resin and each solvent is assigned a value of 1 = dissolved and 0 = not dissolved. This information is feed into a computer minimization program and a solubility envelope is defined. Solvents or solvent blends inside this solubility envelope will remove the soil and those outside will not. This is the basis used to develop cleaning

Solubility Envelope

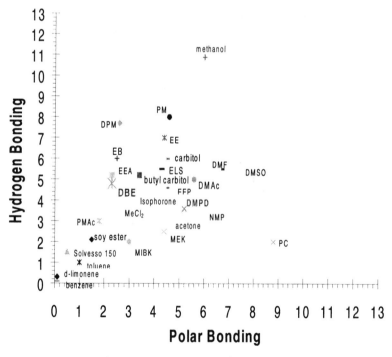

Figure 2 – Hansen solvent map

formulations. An example demonstrating the similarities in solvency between DBE and methylene chloride as compared to other common solvents is shown in Figure 3. Figure 3 shows a solubility envelop for a black enamel paint determined by laboratory testing. The envelope demonstrates that DBE could be used as a replacement for methylene chloride to remove the black paint. However, in contrast to methylene chloride, DBE is worker and environmentally friendly as it is not carcinogenic or an air pollutant. The solvency power combined with the "green" properties of DBE makes it the solvent of choice for the paint stripping industry.

The real power of the Hansen system for the formulator stems from the fact that a simple mixing rule can be applied according to the following equations to derive the solubility parameters of a solvent blend:

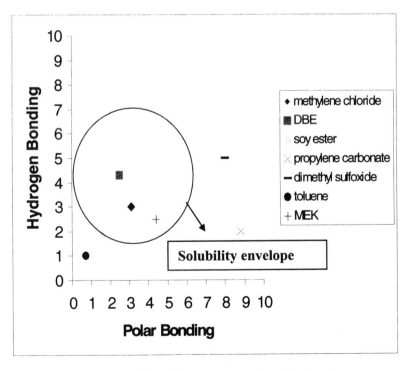

Figure 3 – Solubility envelope for black paint

$$\delta_D = \emptyset 1\ \delta_{D1} + \emptyset 2\ \delta_{D2} + \emptyset 3\ \delta_{D3} + \dots$$

$$\delta_P = \sum \emptyset i\ \delta_{Pi}$$

likewise

$$\delta_H = \sum \emptyset i\ \delta_{Hi}$$

where $\emptyset i$ = weight fraction of the "i"th solvent in a blend.

Selection of low risk solvents

In all applications people desire to work with a low risk solvent for their own safety as well as for the environment. As described previously, in most cases an organic solvent is necessary to meet the needs of the end user. Given this, the question then becomes how does one evaluate and compare solvents so that one of low risk is selected?

To assist in the protection of workers against the health effects of chemical exposure several exposure limits have been established[5]. Exposure limits refer to airborne concentrations of a substance and represent conditions under which it is believed that nearly all workers may be repeatedly exposed day after day without adverse health effects. Exposure limits are not intended for use as a comparison metric of one solvent versus another. However, in practice we have found that in many cases exposure limits are being used to compare the health and safety aspects of using one solvent versus another. Exposure limits are not necessarily an indication of chemical toxicity. They are established to prevent any adverse effect on the health of the user. For example, one chemical's exposure limit of 25 ppm may be established due to carcinogenic concerns and another chemical may have an exposure limit of 25 ppm set due to skin irritation. Therefore, the exposure limits do not reflect the consequences of exceeding them, and many times chemicals with the same exposure limit will have significantly different consequences. Several different agencies have established exposure limits including Occupational Exposure Limits (OEL, PEL) established by OSHA, and Threshold Limit Values (TLV) established by The American Conference of Governmental Industrial Hygienists. DuPont through testing at its Haskell Laboratories sets acceptable exposure limits (AEL), which are equal to or lower than TLV values for a given chemical, and are intended for internal guidance with chemicals.

The current exposure limits set airborne concentration limits on chemical exposure, but do not describe the ease with which that airborne limit can be achieved. For example, if chemical A has a exposure limit of 100 ppm and chemical B has an exposure limit of 5 ppm, many would consider chemical A to be the lower risk solvent. This may or may not be correct since the vapor pressure of the chemicals also must be considered. Vapor pressures describe how much of the chemical is in the vapor phase (airborne) at a particular temperature and pressure. This in a sense determines how much of a chemical a worker will be exposed to through inhalation. The lower the vapor pressure the lower the airborne concentration. The saturation concentration, which is calculated from the vapor pressure, describes how much chemical vapor the air can hold at equilibrium at a given set of conditions. This approach allows one to place a physical description of the fact that at higher temperatures the likelihood of chemical exposure from a solvent increases. Using the Antoine constants (which are tabulated in several reference handbooks for most solvents) one can calculate the vapor pressure of a solvent as a function of temperature and from this calculate a saturation concentration.

Solvent risk is equal to the solvent hazard plus the exposure level. Since the hazard associated with a solvent is inherent in the molecule and can not be changed, then the risk needs to be managed by controlling the exposure. This is an important concept, that the only way to manage the risk of working with a chemical is by managing the exposure. In order to more accurately evaluate solvent exposure both exposure limits and saturation concentration must be considered. We have developed a metric that combines both the exposure limit and vapor pressure in a simple easy to use, easy to understand equation called: the vapor hazard ratio (VHR)[6]. The lower the VHR the lower the risk of achieving the exposure necessary to obtain the health hazard associated with the chemical. The VHR is easily calculated for any chemical from readily accessible data contained in the MSDS by using the following equations:

The saturation concentration of a chemical is calculated by:

Saturation Concentration, ppm = (vapor pressure of chemical, mmHg / 760)*10^6

Then the VHR is:

VHR = (saturation concentration, ppm) / (exposure limit, ppm)

The concept of the vapor hazard ratio is currently being used globally to aid in the evaluation of the potential health hazards of chemicals. Occupational exposure standards are derived and implemented from country to country[7]. Several countries have guidelines that include vapor pressure considerations in addition to exposure limits. In Germany the vapor hazard ratio is described and recommended as "Gefahrdungszahl Gz" in the technical guidelines of the German "Gefahrstoffverordnumg" (TRGS 420, the German technical rule for hazardous materials). The Danish MAL system reflects the cubic meters of fresh air required for ventilation of 1 liter of chemical to below the exposure limit. This number is modified by a constant, depending on the evaporation rate or vapor pressure. Higher vapor pressures imply a greater hazard so the multiplier is larger. Other metrics in addition to exposure limits have been generated to help evaluate risks by inclusion of vapor pressure considerations. A Danish publication comparing several of these is available[8]. The Danish MAL codes define which safety precautions are required for each of a large number of processing operations and conditions. The purpose of such a system is to suggest that possible substitutes to a current solvent can be found with a lower potential health risk. As mentioned previously, in the U.S. it has been our experience that exposure limits are being used as the metric for comparison of one solvent versus another. The additional consideration of the vapor pressure, as done with the VHR, will help solvent users to consider additional metrics in their evaluation.

Although the VHR describes the risk potential based on exposure limits; it fails to describe other hazards which employer and employee may wish to consider to insure the safest workplace possible[9]. A common concern when using solvents is the risk of fire. Every year numerous injuries occur which are the result of a fire involving a solvent[10]. Other possible hazards/factors include chemical incompatibility, reaction with the process or other solvents, and the biodegradability of the solvent. One must also consider that many chemicals have ceiling limits of exposure. Employers/workers, in addition to looking at the VHR should consult the MSDS to determine the human health effects associated with exceeding the exposure limit.

Table 1 contains a list of some of the more common solvents used and their corresponding VHR at room temperature and pressure.

The following example shows how the VHR can be easily and proactively used to develop safer solvent formulations. A cleaning solvent is made up of 40% methyl ethyl ketone (MEK), 40% xylene, and 20% d-limonene. The manufaturer desires to replace the MEK and xylene since as shown in Table 1 they are high-risk solvents.

Table 1: Common Solvents and Their VHR Value

Solvent	TLV/ AEL (ppm)	Vapor pressure (mmHg)	Saturation Conc. (ppm)	VHR
Propylene Carbonate	20	0.02	26	1
DB	5*	0.02	26	5
DPM	100	0.6	723	7
NMP	10*	0.3	394	39
DMSO	10*	0.6	789	79
DBE	1.5*	0.1	131	87
DMAc	10	2	2632	263
Xylene	100	20	26316	263
MIBK	50	15	19737	395
DMF	10	3	3947	395
MEK	200	70	92105	460
Acetone	500	248	326316	652
Toluene	50	28	36842	737
Methanol	200	127	167105	835
Methylene Chloride	50	350	460526	9210
Benzene	1	95	125000	125000

Increasing safety ↑

* DuPont AEL exposure limits were used since no TLV values available.

DB – Di(ethylene glycol) butyl ether
DPM – Di(propylene glycol) methyl ether
NMP – N-methyl-2-pyrrolidone
DMSO – Dimethyl sulfoxide
DBE – Dibasic ester by DuPont
DMAc – N,N-Dimethylacetamide
MIBK – Methyl isobutyl ketone
DMF – N,N-Dimethylformamide
MEK – Methyl ethyl ketone

Solvent	Solvent VHR	Solvent TLV/Exposure limit (ppm)
40% MEK	461	200
40% Xylene	263	100
20% d-limonene	53	50

An exposure limit (TLV or AEL) for a liquid mixture can be calculated on a time-weighted average exposure basis assuming the atmospheric concentration is similar to that of the original mixture (all the liquid mixture eventually evaporates). This approach is not valid if the health end effects of the individual solvents differ dramatically. When the percent composition (by weight) is known the exposure limit of the mixture is:

Exposure limit mixture = 1/[(fa/TLVa) + (fb/TLVb) + (fc/TLVc) +...(fn/TLVn)]

Where fn is the weight fraction of component n and TLVn is the exposure limit of component n. An alternative method is using the VHR in place of the TLV/exposure limit to reflect the health risk a solvent mixture poses based on the not only the exposure limit but the potential ease at which that limit could be achieved at a given set of conditions:

Exposure limit mixture = 1/[(fa/VHRa) + (fb/VHRb) + (fc/VHRc) +...(fn/VHRn)]

Using a solvent formulation program (DuPont as well as many other companies offer such services to customers), the Hansen parameters for the above hypothetical solvent blend is found to be non-polar 9.1, polar 2.3, and hydrogen bonding 1.7. The Hansen parameters show the solvency of the solvent blend and it is desired that a replacement solvent blend has similar Hansen parameters and present less of a health risk. As is almost always the case there is not one single correct answer for a replacement solvent in a blend, rather many are possible. A few possible answers could be:

Blend #	Solvents	Solvent VHR	Solvents TLV/Exposure limits (ppm)
Blend 1	40% d-limonene 35% DuPont DBE 10% propylene carbonate 15% NMP	53 87 1 16	50 1.5 20 25
Blend 2	40% DuPont DBE 20% NMP 40% Solvesso 150	87 16 26	1.5 25 50
Blend 3	10% DMSO 30% EB 60% mineral spirits	79 39 39	10 20 100

The Hansen parameters for the proposed solvent blends 1-3 have similar and acceptable solvency to the current blend, as shown in Figure 4. Comparison of the VHR values of the individual solvents indicates that the proposed formulations are solvent blends of lower hazard potential than the current blend. The VHR are TLV/exposure limits for the blends were calculated:

Solvent Blend	Blend exposure limit (using VHR)	Blend exposure limit (using TLV)
Current Blend	163	100
Blend 1	9	4
Blend 2	31	3
Blend 3	41	12

The calculations show that by inspection of only the TLV/exposure limits of the blends (the lower the TLV the more hazardous the blend) one may conclude that the current blend has the lowest potential health risk. However, by calculating the VHR of the blends (the higher the VHR the more hazardous the blend) one

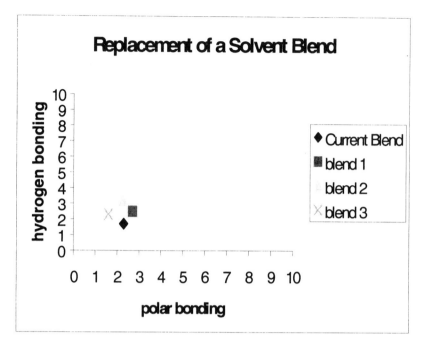

Figure 4 – Comparison of replacement blends

may conclude that the current blend has the highest potential health risk to the user. This example highlights the fact that a different conclusion on the relative safety of using one solvent blend versus another can be reached depending on how the evaluation method used. In this example the VHR regards the current blend as being more hazardous due to the fact the vapor pressure shows it to be more volatile, and hence its exposure limit may be more easily achieved. Since the exposure limit alone does not reflect the potential volatility of a solvent blend that information is not considered if one compares solvent blends using only exposure limits. Clearly, the type of information provided by the VHR is of value since it provides another approach to aid in the evaluation of the relative hazards of using one solvent blend versus another.

Conclusion

Solvents have been and will continue to be needed and used by industry and consumers. There is an increased awareness around people and environmental safety when working with solvents. This has led to the desire for the use of low risk green solvents in applications where solvents are used. Dibasic ester, DBE is an example of a low risk green solvent that is produced from a plant co-product stream. An easy to use metric, VHR, the vapor hazard ratio is presented to aid in the selection of low risk solvents.

References

1. Futuretech "solvents for green chemistry", Wiley Press Oct. 11, 1999, no.242

2. Kirk Othmer Encyclopedia of Chemical Technology, vol. 22, 1998.

3. Dibasic Ester Solvent Brochure, 1999.

4. Charles Hansen, "Hansen Solubility Parameters", CRC Press 2000

5. American Council of Governmental Industrial Hygiensts, "threshold limit values", 1990

6. Kob, N.; Altnau G. CleanTech magazine, May 2000.

7. Jankovic, J., Drake F., Am. Ind. Hyg. Assoc. J. **57**, (1996), 641

8. Vincent, J., Am. Ind. Hyg. Assoc. J. **59**, (1998), 729

9. Filskov, P.; Goldschmidt, G.; Hansen, M.F.; Hoglund, L.; Johansen, T.; Pederson, C.L.; Wibroe, L. *Substitution in practice – experience from BST* Arbejsmiljofondet, Copenhagen, 1989.

10. Straughn, J., *Chemical Health & Safety*, May/June 1999, 33.

INDEXES

Author Index

Subject Index

A